WILD EDIBLE FUNGI

A GLOBAL OVERVIEW OF THEIR USE AND IMPORTANCE TO PEOPLE

WILD EDIBLE FUNGI

A GLOBAL OVERVIEW OF THEIR USE AND IMPORTANCE TO PEOPLE

Eric Boa

Reprinted by arrangement with the
Food and Agriculture Organization of the United Nations

by
Daya Publishing House
A Division of
Astral International Pvt Ltd
New Delhi - 110 002

"This book was originally published by the Food and Agriculture Organization of the United Nations (FAO) as Vaccine Manual, FAO Animal Production and Health Series No. 35".

First Indian Edition, 2007
Reprinted, 2024
ISBN: 978-81-70354-98-7

Published by : **Daya Publishing House**
A Division of
Astral International Pvt Ltd
4736/23, Ansari Road
Darya Ganj, New Delhi - 110002
Phone: 011-43549197
Email: info@astralint.com
Web: www.astralint.com

Printed at : **Chawla Offset Printers,**
Delhi-110 052

TABLES

PLATES

FIGURES

BOXES

FOREWORD

Much of the forestry and development debate in international fora focuses on how forests, forest products and forestry can contribute to the United Nations Millennium Development Goals of halving poverty and food insecurity by 2015. FAO's programme on the Promotion and development of non-wood forest products (NWFP) is contributing to this goal by improving the sustainable use of NWFP in order to improve income-generation and food security, to contribute to the wise management of the world's forests and to conserve their biodiversity.

One of the important groups of NWFP, collected all over the world and used for subsistence purposes as well as sold on local markets and restaurants, are fungi – often called mushrooms. However, most information on fungi is available on cultivated species while data on wild edible fungi (WEF) remain scarce.

The present publication was supported by funds from the Norway Partnership Programme "Forests for sustainable livelihoods". It aims at documenting and analysing the role of WEF in food security with focus on developing countries. It compiles in one volume the much scattered information on the uses and prospects for development of WEF, including issues such as ecology, collection, harvesting, socio-economic benefits and trade.

By disseminating this information, it is expected that the attention of forestry technicians, nutritionists, natural resource planners, policy-makers and other stakeholders concerned will be drawn to the links between this important category of NWFP, food security and sustainable forest management.

It is hoped that the use of this document will help to promote the sustainable use of WEF as a valuable component in the process of economic development and poverty alleviation.

Wulf Killmann
Director
Forest Products and Economics Division
Forestry Department

ABBREVIATIONS

DFID	Department for International Development
ECM	Ectomycorrhiza
FAO	Food and Agriculture Organization of the United Nations
NGO	Non-governmental organization
NTFP	Non-timber forest products
NWFP	Non-wood forest products
TCM	Traditional Chinese Medicine
SEMARNAT	Secretariat de Medio Ambiente y Recursos Naturales (Secretariat of Environment and Natural Resources)
WEF	Wild edible fungi
WUF	Wild useful fungi (including those with edible, medicinal and other properties)

ACKNOWLEDGEMENTS

This publication is based on a draft presented by Eric Boa from CABI *Bioscience*. The author first became involved with wild edible fungi through Jim Waller, a colleague from CABI *Bioscience*. This led to the Miombo Edible Fungi project, funded by the Department for International Development (DFID) from 1999 to 2002 and carried out in conjunction with the Forestry Research Institute of Malawi. Paul Kirk, Gerald Meke and Janet Lowore made major contributions to this project and towards the author's own understanding of wild edible fungi and their use by rural people.

The author was repeatedly intrigued and astounded by how much has been written about wild edible fungi, often buried away in reports and other less visible places. Two British colleagues, Brian Morris and Graham Piearce, have been a particular inspiration. Both have undertaken broad and sustained researches that have not only helped to demonstrate the importance of wild edible fungi to people in southern Africa, but have also raised awareness on a broader front of an often neglected resource. Graham has been a generous and prompt provider of comments, information and photographs.

Dave Pilz of the United States is one of a group of scientists who have worked on wild edible fungi in the Pacific northwest. Their work has also been an inspiration and source of insights. The author thanks Dave in particular for the benefit of his insights on critical scientific issues. Jeffery Bentley has done much to educate me on people issues and without this I would still be struggling to make the enriching connections between science and development. At FAO, Laura Russo suggested that the author should write this book, while Sven Walter has overseen the execution, casting a patient yet critical eye over the manuscript. The author thanks him and his colleagues, in particular Florance Egal, François N'Deckere-Ziangba, Jorike Potters, Mette Loyche-Wilkie, Michel Laverdière, Olman Serrano, Paul Vantomme, Peter Bailey and Tina Etherington, for their comments. The author thanks his family for giving him the time to complete the task and other friends who provided him with accommodation and peace to write.

The other people who have helped are arranged alphabetically by first name.

Alessandra Zambonelli (Italy); Anabela Martins (Portugal); Ana Franco-Molano (Colombia); André de Meijer (Brazil); André de Kesel (Belgium); Andrew Pulford (UK); Antonella Amicucci (Italy); Anxious Masuka (Zimbabwe); Arailde Fuentes (Brazil); Arie Bijl (South Africa); Attila Hegedus (Hungary); Bart Buyck (France); Brenda Down (UK and Sierra Leone); Brian Morris (UK); Caroline Boa (UK); Daniel Winkler (USA); Dave Pilz (USA); David Minter (UK); Dennis Desjardin (USA); Duncan Boa (UK); Elaine Marshall (UK and Mexico); Else Vellinga (USA); Emidio Borghi (Italy); Frank Taylor (Botswana); Gene Yetter (USA); Georges René (Italy and Haiti); Gerald Meke (Malawi); Gerritt Marais (South Africa); Gianluigi Gregori (Italy); H Hosaka (Lao People's Democratic Republic); Graham Piearce (UK); Giuseppe Cardinale (Italy); Harry Evans (UK); Hildegun Flom (Norway); Ian Hall (New Zealand); Ibu Leyulani (Indonesia); Irina Gorbunova (Russia Federation); Irma Gamundí (Argentina); Janet Probyn (Lowore) (Malawi); Javier Lander (Spain); Jerry Cooper (New Zealand); Jim Waller (UK); Jimmy Lowore (deceased: Malawi); Jolanda Roux (South Africa); Lauro Russo (Italy); Lesley Ragab (UK); Luba Nanaguylan (Armenia); Luo Ghuozhong (China); M. Snowarski (Poland); Marc Ducousso (France); Maria Chamberlain (UK); Maria Teresa Schifino-Wittmann (Brazil); Marisela Zamora-Martínez (Mexico); Marja Härkönen (Finland); Mary Apetorgbor (Ghana); Meral Gurer (Turkey); Mike Howard (South Africa); Miriam de Román (Spain); Myles Mander (South Africa); Necla Caglarirmark (Turkey); Paul House (Honduras and UK); Paul Kirk (UK); Phuntsho Namgyel

(Bhutan); Pierluigi and Luna the dog (Urbino, Italy); Roberto Flores (Guatemala); Rory McBurney (UK); Roy Watling (Scotland); Sara Maltoni and her mother (Sardinia); Seona Anderson (UK); Sinclair Tedder (Canada); Solomon Wasser (Israel); Soulemane Yorou (Benin); Stephanos Diamandis (Greece); Susan Alexander (USA); Sven Walter (Italy); Tim Livesey (UK); Warren Priest (UK); Yun Wang (China and New Zealand).

These friends and colleagues have provided the author with much information. Any errors in presentation or interpretation remain with the author and FAO. Paul Kirk has checked scientific names for wild fungi and suggested changes to reflect current taxonomic thinking. This is an area of some confusion and there are doubtless further changes that need to be made to the lists compiled for the book. New initiatives on wild edible fungi are being developed, aimed at sustainable use by rural people, particularly but not exclusively in developing countries.

SUMMARY

Wild edible fungi are collected for food and to earn money in more than 80 countries. There is a huge diversity of different types, from truffles to milk-caps, chanterelles to termite mushrooms, with more than 1 100 species recorded during the preparation of this book. A small group of species are of economic importance in terms of exports, but the wider significance of wild edible fungi lies with their extensive subsistence uses in developing countries. They provide a notable contribution to diet in central and southern Africa during the months of the year when the supply of food is often perilously low. Elsewhere they are a valued and valuable addition to diets of rural people.

Commercial harvesting is an important business in countries such as Zimbabwe, Turkey, Poland, the United States of America, the Democratic People's Republic of Korea and Bhutan. The export trade is driven by a strong and expanding demand from Europe and Japan and is predominantly from poor to rich countries. This is good for local businesses and collectors, providing important cash income that pays for children to go to school and helps to reduce poverty in areas where the options for earning money are limited. Local markets around the world reveal a widespread though smaller individual trade in an extensive range of species. Though difficult to measure compared with the more visible export of wild edible fungi, local trade is of considerable value to collectors and increases the supply of food to many areas of weak food security.

Collection and consumption within countries varies from the extensive and intensive patterns of China to more restricted use by indigenous people in South America. Substantial quantities are eaten through personal collections that may go unrecorded. The nutritional value of wild edible fungi should not be underestimated: they are of comparable value with many vegetables and in notable cases have a higher food value.

Wild edible fungi play an important ecological role. Many of the leading species live symbiotically with trees and this mycorrhizal association sustains the growth of native forests and commercial plantations in temperate and tropical zones. The saprobic wild edible fungi, though less important in terms of volumes collected and money earned from local sales, are important in nutrient recycling. The saprobic species are the basis for the hugely valuable global business in cultivated mushrooms, currently valued at around US$23 billion each year. This is an increasing source of income for small-scale enterprises in developing countries.

Wild edible fungi are among the most valuable NWFP with much potential for expansion of trade, but there are also challenges in the integration of their management and sustainable production as part of multiple use forests. There are concerns about the impact of excessive harvesting, which require better data on yields and productivity and a closer examination of collectors and local practices. Closer cooperation between forest managers and those using wild edible fungi is needed and suggestions are made on how this might be achieved.

There is a strong emphasis on subsistence uses of wild edible fungi and their importance to rural people in developing countries, although this is an area where there are still significant gaps in information. There is also significant commercial harvesting in developed countries, such as the United States of America and Canada, and in the emerging economies of eastern Europe, for example Poland and Serbia and Montenegro. However, countries in the north are of greater significance to wild edible fungi as a destination for exports and as a source of scientific expertise, especially in mycology (the study of fungi).

This scientific expertise is increasingly being applied to help achieve the major development goals, which include poverty alleviation and sustainable use of natural resources. Real progress has been and continues to be made in the roles that wild edible fungi contribute towards these goals.

1 Introduction: setting the scene

GENERAL IMPORTANCE

Wild edible fungi (WEF[1]) have been collected and consumed by people for thousands of years. The archaeological record reveals edible species associated with people living 13 000 years ago in Chile (Rojas and Mansur, 1995) but it is in China where the eating of wild fungi is first reliably noted, several hundred years before the birth of Christ (Aaronson, 2000). Edible fungi were collected from forests in ancient Greek and Roman times and highly valued, though more by high-ranking people than by peasants (Buller, 1914). Caesar's mushroom (*Amanita caesarea*) is a reminder of an ancient tradition that still exists in many parts of Italy, embracing a diversity of edible species dominated today by truffles (*Tuber* spp.) and *porcini* (*Boletus edulis*).

China features prominently in the early and later historical record of wild edible fungi. The Chinese have for centuries valued many species, not only for nutrition and taste but also for their healing properties. These values and traditions are as strong today as they were centuries ago and are confirmed by the huge range of wild fungi collected from forests and fields and marketed widely (Wang, 1987) (Plate 8). China is also the leading exporter of cultivated mushrooms.

It is less well known that countries such as Mexico (Plate 7) and Turkey, and major areas of central and southern Africa (Plate 6), also have a long and notable tradition of wild edible fungi. The list of countries where wild fungi are reported to be consumed and provide income to rural people is impressive (Annex 1).

The threat posed by poisonous and lethal species is often overstated. Incidents of poisoning and deaths are few and far between compared to the regular and safe consumption of edible species, but publicity and cultural attitudes continue to fuel an intrinsic fear of wild fungi in some societies. This is more commonly found in developed countries and has undoubtedly led to general beliefs that global use of wild edible fungi is small-scale and restricted to key areas. As this publication conclusively shows, this is simply not true (Table 1). The use of wild edible fungi is both extensive and intensive, though patterns of use do vary (Annex 1).

Wild edible fungi add flavour to bland staple foods but they are also valuable foods in their own right. Local names for termite mushrooms (*Termitomyces*) (Plate 6) reflect local beliefs that they are a fair substitute for meat, a belief that is confirmed by nutritional analyses. Not all wild edible fungi have such a high protein content but they are of comparable nutritional value to many vegetables.

In addition to making substantial contributions to the diets of poor people in developing countries, they are an important

TABLE 1
Numbers of species of wild edible and medicinal fungi

CATEGORY	NO. OF SPECIES	PERCENTAGE TOTAL
1. Edible only	1 009	43
2. Edible and medicinal	88	4
3. Food only	820	35
4. Food and medicinal	249	11
5. Medicinal only	133	6
6. Other uses (none of above)	29	1
TOTAL wild useful species	**2 327**	
ALL edible only (1+2)	1 097	
ALL food (3+4)	1 069	
ALL medicinal (2+4+5)	470	

Note: Compiled from more than 200 different sources from 110 countries, but excludes a detailed review of species from developed countries. Varieties and subspecies are counted separately. The categories *food* and *edible* are mutually exclusive. To distinguish clearly between use and properties of a species: substantial numbers of edible species lack confirmed use as food.

[1] See Box 1 for a discussion of terminology used in this book.

BOX 1
Wild edible fungi and mushrooms

Fungi are a distinct group of organisms which include species with large and visible fruiting bodies (macrofungi). The best known examples of macrofungi are the mushrooms. They have a cap and a stalk and are frequently seen in fields and forests. Most are simply inedible but there are notable examples that can be eaten. The number of poisonous species is relatively small while those that are fatal belong to a tiny minority. The most familiar edible mushrooms are those that are cultivated and sold fresh and tinned in shops.

Macrofungi have many different shapes and appearances. Boletes have pores rather than gills on the underside of the cap; truffles grow underground and do not have a stalk and a cap (Plate 1). *Huitlacoche* is a Mexican food produced when maize cobs are infected by a fungus. This is clearly not a mushroom.

Wild edible fungus (fungi is the plural form, usually pronounced with a hard "g") is used to distinguish their origin and the fact that they include a variety of forms that include infected maize cobs, stomach fungi, boletes, bracket fungi and, of course, mushrooms. Many other publications (e.g. Hall *et al.*, 1998a) refer to wild mushrooms, defining this broadly to include the different shapes and appearances.

It is interesting to compare terms used in other languages. In Italy wild fungi are referred to as *funghi comestible*; there is no equivalent of "mushroom" in Italian. In Spanish *hongo comestible* and *hongo silvestre* are used. *Seta* is similar in meaning to mushroom but it does not imply that a particular species is edible. In Malawi, *bowa* describes an edible fungus in the Chewa language, a term that has in essence the same meaning as "wild edible fungus".

source of income. Wild edible fungi are sold in many local markets and commercial harvesting has provided new sources of income for many rural people. The demand for specialist wild mushrooms from Europe and Japan continues to earn significant amounts for countries such as Bhutan, the Democratic People's Republic of Korea and Pakistan.

Wild fungi also have medicinal properties, some of which are found in edible species (Table 1). Wild *useful* fungi therefore contribute towards diet, income and human health. Many species also play a vital ecological role through the symbiotic relationships known as mycorrhizas that they form with trees. Truffles and other valuable wild edible fungi depend on trees for their growth and cannot be cultivated artificially. The mycorrhizas enable trees to grow in nutrient-poor soils. The trees of the miombo woodland of central and southern Africa and the woodland itself would not exist without their fungal partners.

The importance of wild edible fungi continues to grow for more fundamental reasons. Logging bans in several countries has renewed interest in non-wood forest products (NWFP) as an alternative source of income and jobs for people previously employed in forestry. Wild edible fungi have played an important role in providing new sources of income in China and the United States of America. Further information is given in Chapters 3 and 4.

To summarize, wild edible fungi are important for three main reasons:

- as a source of food (plus health benefits);
- as a source of income;
- to maintain the health of forests.

TRADITIONS AND HISTORY OF USE

Ethnomycology is the study of people and fungi and is a recent area of academic interest. It traces its roots to a landmark publication entitled *Mushrooms, Russia and history* (Wasson and Wasson, 1957). This privately published and discursive opus contains a wealth of useful information on the culture and history of wild fungi. Although ethnomycology began with a clear interest in wild edible fungi, later developments saw a strong emphasis on hallucinogenic mushrooms and their cultural significance (Schultes, 1940; Wasson, 1968). While this continues to be an area of understandable intrigue, the spotlight has turned back to wild edible fungi. During the last twenty to thirty years researchers have substantially increased our knowledge of local traditions in Africa, Asia and Mesoamerica (Mexico, Guatemala).

An early distinction was made between *mycophilia* and *mycophobia*: In *mycophilic* societies or cultures, fungi are esteemed and there is a strong and long tradition of popular use. *Mycophobic* cultures have a minor regard for fungi and they are often actively feared (Wasson and Wasson, 1957). The British colonial record in Africa contains little information about the local use of wild edible fungi, despite the fact that people throughout southern Africa have eaten them for centuries (Morris, 1994; Piearce, 1985).

The history of use of wild edible fungi is well recorded for China, although much information is still in Chinese (Plate 3). China is an example of a mycophilic country while Britain is usually classified as mycophobic. These distinctions are becoming less clear, however, and although there is still a weak tradition of collecting in northern Europe in general, more people gather from the wild than before. Some of this is for commercial reasons (Dyke and Newton, 1999) but immigrants from mycophilic countries have also changed attitudes. There is an expanding group of people who now collect wild edible fungi in the United States, for example. Traditions vary within countries: the central and northern regions of Italy are strongly mycophilic, but the tradition of collecting and consuming wild edible fungi is less strong in the south. Catalonia in Spain has a markedly greater interest in wild edible fungi compared to other regions. Variable traditions also exist in the United Republic of Tanzania (Härkönen, Saarimäki and Mwasumbi, 1994).

Finland represents a particularly interesting meeting of traditions. The historical influence of Swedish culture did little to enthuse local interest in the west but, in the east, Karelian people who came from the Russian Federation to live in Finland brought a much stronger tradition and liking for wild edible fungi (Härkönen, 1998). Russians are noted for their general (though not universal) love of wild edible fungi, as witnessed by large-scale movement of people to forests at the weekend (Filipov, 1998). The Estonians have a saying that describes the Russian passion: "Where there is a mushroom coming up, there is always a Russian waiting for it". In Finnish Karelia they used to say "Shouting like Russians in [a] mushroom forest"[2].

The Latin American tradition is almost wholly restricted to Mexico (see review by Villarreal and Perez-Moreno, 1989). It extends south to Guatemala and briefly into Honduras (House, 2002, personal communication: *Wild edible fungi in Honduras*) but then abruptly ends, despite the widespread occurrence of pine forests and other trees with edible mycorrhizal fungi. There is little evidence of strong traditions in South America, although studies of native people in Amazonia (Prance, 1984) revealed regular consumption and management of wild edible fungi (though all saprobic). A little-known study from Papua New Guinea (Sillitoe, 1995) reveals a wealth of information on wild edible fungi that hints at wider use in other countries.

Local people reject some species that are edible. Boletes are not eaten in parts of the United Republic of Tanzania as a general rule (Härkönen, 2002). An Italian priest living in Guatemala found that local people were ignoring *Boletus edulis*, despite their general liking of wild edible fungi. With his encouragement they were able to enjoy a species they had previously ignored (Flores, 2002, personal communication: *Guatemala edible fungi*). It is not clear whether people in Europe would readily eat *Phallus impudicus*, however, despite its widespread popularity in China (Plate 9) and some cultures are

[2] Information provided by Marja Härkönen.

BOX 2

A developing country perspective

Most of the information on the biology and ecology of edible macrofungi is based on research carried out in developed countries. The literature is heavily weighted towards perceptions of value and usefulness of wild edible fungi found in the North. Here there is a strong emphasis on valuable types such as true truffles (*Tuber* spp.), chanterelles and various boletes – of which *Boletus edulis* is the best known. There is much less knowledge, for example, about the many species of *Lactarius* or *Russula* eaten in Africa, from a biological, social or economic perspective.

Income from wild edible fungi is an important source of revenue for rural communities, especially in developing countries. In central southern Africa, WEF are a significant source of nutrition; so too in rural parts of China, India and Mexico. In Europe, WEF are a specialist food, a gourmet item to be savoured infrequently – a reflection of the high prices demanded for prized species. This can mean good incomes for the less well-off in rural parts of Spain and Italy, but the overall importance of WEF to such societies, and indeed the potential for increased local incomes, is small compared to local use and markets in the developing world.

The publication concentrates on improving knowledge about wild edible fungi in developing countries, though research and published information from the North has not been ignored. The experiences in the Pacific northwest of North America have been widely quoted following an expansion of activities on NWFP generally as traditional forestry industries declined and rural communities sought new sources of income. Finland has long promoted a wider use of wild edible fungi as the country emerged from times of economic difficulty, while the demand for *matsutake* (*Tricholoma* spp.) in Japan has been of major significance for developing countries such as China, the Democratic People's Republic of Korea and even Bhutan.

These examples offer wider insights on a number of different aspects of wild edible fungi, from management of natural resources to collection practices. A comprehensive review of WEF use in the South and in the North is, however, beyond the scope of the current publication. That is not to say that the collection of truffles in Italy or France, or *níscalos* (*Lactarius deliciosus*) in Spain, are without economic importance to local people (de Román, 2002, personal communication: *Trade in níscalos from North Spain to Catalonia and truffle production*), but again these are of lesser significance as a source of income compared with comparable activities in many developing countries.

instinctively fearful of *Amanita* species. This genus contains deadly poisonous as well as flavoursome species (Plate 7).

PURPOSE AND STRUCTURE OF THE PUBLICATION

This publication presents information on the importance of wild edible fungi to people. It includes details of species collected and sold, but there is a particular emphasis on social and economic aspects in an attempt to show how wild edible fungi might contribute to rural livelihoods. There is a wealth of information on the biology and general characteristics of macrofungi but this is only discussed in detail where it is relevant to either people or the use of natural resources.

A broader aim of the publication is to increase awareness of wild edible fungi and to emphasize the ecological link between key species of wild edible fungi and forests. Suggestions are made on how to manage wild edible fungi in a sustainable manner, taking into account the multiple use of forests and other forest users.

There is a strong emphasis on developing countries (the "South") in many of the discussions, particularly when reviewing how to improve the benefits of wild edible fungi and their sustainable production. Information is also drawn from case studies and experiences with wild edible fungi in developed countries (the "North"). The reasons for concentrating on developing countries are explained in more detail in Box 2.

The publication is divided into five chapters and includes a comprehensive series of tables and annexes. The reader is pointed towards primary and other sources of information, bearing in mind that personal communications with the authors have been an important means of learning more about wild edible fungi. Original publications are often difficult to obtain and general sources such as the excellent literature reviews by Rammeloo and Walleyn (1993) and Walleyn and Rammeloo (1994) are recommended for Africa south of the Sahara. There is a need to undertake similar reviews for other regions of the world, thus increasing awareness of a surprising breadth of published information and stimulating interest in new lines of research.

The characteristics of wild edible fungi are briefly described in Chapter 2 and include key facts on biology, ecology, edibility and cultivation. The emphasis is on general information and few technical details are presented. Major genera are described in outline. Latin names are mostly used throughout the book since there are few common names for wild edible fungi that easily transfer from one country or language to another. The exceptions include *matsutake* (*Tricholoma matsutake* but also *T. magnivelare* and other species), chanterelles (*Cantharellus* spp.) and *porcini* (*Boletus edulis*).

Management issues are explored in Chapter 3, and this includes a review of collectors and the relationship between harvesting and forest management. This section examines some of the broader issues concerning multiple use of forests, productivity of wild edible fungi and access to collecting sites.

Chapter 4 continues the discussions begun in the previous chapter but pays particular attention to people and how wild edible fungi are traded, their importance to diet and nutrition. Trade data, marketing and commercialization are explored, with a final section that briefly summarizes the use of wild edible fungi by region and country.

The final chapter moves from the present to the future. Chapter 5 examines possibilities for future initiatives with an emphasis on practical steps that could be taken to increase benefits to rural people while sustaining production of wild edible fungi and achieving sound forest management. The publication describes known constraints to the sustainable use of wild edible fungi. A forest manager in western China despaired of getting local collectors to adopt less destructive harvesting practices for a prized edible fungus known as *matsutake* (Winkler, 2002). Such constraints, it is suggested, can be overcome and changes effected, but only if actions are based on a sound knowledge of what people do and why.

SOURCES OF INFORMATION

Information has been gathered on activities in 85 predominantly, but not exclusively, developing countries (see Box 2). The published and accessible information has varied hugely in terms of emphasis (Table 2), detail and accuracy and has demanded careful examination. There are many mycological publications, for example, that list species as being "edible" but do not elaborate on their local use. General accounts of NWFP talk about "mushrooms" without specifying which types.

Over 800 papers, books, newspaper articles, personal communications, Web sites and miscellaneous other sources of information were consulted. Information on wild edible fungi is spread across many different disciplines (Table 2). Each discipline represents a different set of interests but also with some overlap. This is the first time that a broad review of wild edible fungi has been attempted and there is undoubtedly still much to learn, particularly from the Russian and Chinese literature. Information about wild edible fungi in the Russian Federation was only available because of a translation made by Dr Seona Anderson of a key text (Vasil'eva, 1978).

TABLE 2
Disciplines and areas of activity containing information on wild useful fungi

Discipline or topic	Notes
Mycology, including mycorrhizas	The study of fungi (mycology) includes molecular biology, biochemistry and more traditional topics such as ecology and taxonomy. Published information generally has little detail about the use of fungi by people, particularly the social and economic aspects. Mycorrhizal studies have a combined interest in fungi and plants. Edible ectomycorrhizal fungi have only recently emerged as a subdiscipline within a much larger area of study.
Field biology and natural history	Field guides contain descriptions of species and photographs and are used mostly for identification purposes. The majority of guides are published in the North and therefore have a limited use in developing countries. A few guides are specifically for edible fungi. Natural history publications have provided some information on uses of WEF by people, though this group is often ignored or dealt with fleetingly.
Cultivation of mushrooms	There is an extensive literature on cultivated mushrooms. Regular meetings are held which have strong commercial support. There has been recent interest (e.g. Mshigeni and Chang, 2000) in the introduction of small-scale production units to developing countries and a small but growing literature on managing natural areas for production of *matsutake* and truffles (Federation-Francaise-des-Trufficulteurs, 2001).
Ethnomycology	Ethnomycology is a relatively young area of investigation. Topics include the cultural, ceremonial and medicinal uses of fungi by people. Ethnomycology was originally dominated by the study of hallucinogenic mushrooms and their cultural significance and little attention has been paid to the uses of WEF by people.
Nutrition, human health, food security	The literature on nutritional value is surprisingly large though analytical approaches vary and comparison of results is difficult. Most analyses have involved cultivated mushroom species with only a few wild edible species included. There has been a huge expansion of scientific research on cultivated, medicinal mushrooms, mushrooms as dietary supplements and "nutriceuticals", but this is of limited relevance to development initiatives. There are few studies that have considered wild edible fungi in the context of food security, though this angle deserves closer attention.
Markets and trade	Data on volumes and values of wild edible fungi collected are weak, patchy and often unreliable. Global estimates of trade are open to interpretation and unreliable sources may acquire a spurious credibility by repeat references. Although caution is needed when reviewing marketing data there has been more accurate documentation in recent years.
Wood and non-wood forest products	Wild edible fungi appear regularly in NWFP studies but individual species are often not mentioned (if ever identified). Specific and detailed interest has grown as the result of activities in the Pacific northwest of the United States and Canada and elsewhere. General NWFP studies are often a disappointing source of information on wild edible fungi.

2 Characteristics: biology, ecology, uses, cultivation

Mycology is the study of fungi and mycologists are the people who carry out these studies. New research methods have substantially increased knowledge about the fundamental nature of fungi. Much of this research has focused on fungi that cause plant diseases. Research on edible fungi has concentrated on a small group of species that are commercially cultivated. Wild edible fungi have, until recently, been relatively ignored by science, though amateur mycologists often documented species they found in field studies, mostly in Europe or countries in which Europeans have settled.

There has always, however, been a keen interest in a small group of valuable wild edible fungi that cannot be cultivated. These include the truffles (*Tuber* spp.), *matsutake* (*Tricholoma* spp.) and *porcini* or *cèpes* (*Boletus edulis*). Their biology and ecology have been studied in some detail – a marked contrast to the many other wild edible fungi used around the world.

The consequence of this neglect is that wild edible species used in developing countries are poorly known. Some information is available from studies of close relatives in temperate regions. *Russula* and *Lactarius* occur around the world, for example, and knowledge of species in Europe can be applied with some caution and caveats to African species. The main problem is naming and recognizing species. Genera and species concepts were originally based on the narrower range of diversity found in temperate regions and these may require fundamental reappraisal as tropical species become better known.

This chapter provides a brief introduction to the larger fungi (macrofungi), with special reference to those that are edible. The use of specialist terms has been avoided where simpler alternatives are available. Field guides contain useful glossaries and there are an increasing number of Web sites that help in understanding technical terms (Chapter 6). The *Dictionary of the fungi* is a regularly updated text with details about all fungal genera and other information on mycology (Kirk *et al.*, 2001).

WHAT ARE FUNGI?

Fungi are a distinct group of organisms more closely related to animals than plants. At present fungi are divided into three separate and distinct kingdoms based on an expanded knowledge of their biochemistry and genetic makeup established especially over the last 30 or so years. It is wrong and misleading to refer to fungi as "plants without chlorophyll" (FAO, 1998a).

Despite fundamental differences, fungi are often classified as plants. Understanding the taxonomic status of fungi has little apparent significance to people collecting and selling wild edible fungi, but it is of critical importance in establishing a sound and robust classification system. This ensures that when two people use the same species name they know that they are referring to the same (edible) fungus.

The classification of fungi with plants has inadvertent practical consequences. It is not always clear whether ethnobotanical studies include wild fungi, as is the case with a study from Turkey (Ertrug, 2000). **Ethnomycology** is the correct term that indicates fungi are involved. On a similar track, flora refers only to plants. The equivalent term for fungi is **mycota.** These fungal terms may be unfamiliar but their use helps to identify published information on wild edible fungi clearly that may otherwise be ignored or missed.

Structure and feeding

Fungi come in many shapes, sizes and colours (Plate 1). Macrofungus (plural: macrofungi) is a general category used for species that have a visible (to the unaided eye) structure that produces spores, such as a mushroom or truffle. These visible structures are generically referred to as "fruiting bodies".

Fungi consist of fine threads known as hyphae, which together form a mycelium, as in the mould growing on a piece of fruit or bread. The cap of a mushroom or a bracket fungus also consists of hyphae, densely packed together to form the fruiting body. Specialized hyphae produce spores that are dispersed in a number of ways. They can be viewed en masse by placing the cap of a mushroom on a piece of white paper and covering it with a glass (Plate 3). The colour, form and way in which spores develop help to identify the fungus.

Wild edible fungi are often referred to generically as wild edible "mushrooms". This can be confusing for a number of reasons: edible species have different forms, some with gills and some with pores, some with stems and some without (Plate 1). This book prefers the broader term wild edible fungi to reflect the diversity of forms and also to distinguish them clearly from cultivated mushrooms (Box 1).

How fungi feed

Fungi are dependent on dead and living material for their growth. They obtain their nutrients in three basic ways:

- SAPROBIC[3] – growing on dead organic matter;
- SYMBIOTIC – growing in association with other organisms;
- PATHOGENIC or PARASITIC – causing harm to another organism.

The majority of wild edible fungi species are symbiotic and form mycorrhizas with trees (see below). Saprobic edible fungi are also collected from the wild but they are best known and most widely valued in their cultivated forms. Plant pathogenic fungi cause diseases of plants and a small number of these microfungi are eaten in the form of infected host material (Plate 2). The different modes of feeding are shown in Plate 2 and described briefly below.

Saprobic fungi

Fungi colonize rotting wood and organic matter found in soil. Many species cannot be seen with the naked eye (microfungi) but there are (edible) macrofungi that fruit on fallen logs and bracket fungi that grow from dead or dying parts of standing trees. *Agaricus arvensis* is a commonly collected wild edible species that occurs in pastures and grassy areas. Edible species of *Favolus* are collected from dead wood inside tropical rain forests. The wild edible fungi used by the Yanomam Indians in Brazil are all saprobic and occurred in slash and burn areas where rotting wood was present (Prance, 1984).

In the wild, the volume and value of saprobic species used as food are small by comparison with the symbiotic edible fungi, though more edible saprobic species are collected. Their overall value is much higher because they are widely cultivated: a recent figure of US$18 billion was quoted for the annual, global trade in cultivated, saprobic species (Chang, 1999; see also Table 19).

Saprobic species need a constant supply of suitable organic matter to sustain production in the wild and this can be a limiting factor in production. *Shi'itake* (*Lentinula edodes*) mushroom cultivation in one area of China is threatened by the supply of suitable tree branches from nearby forests (Pauli, 1998).

Saprobic macrofungi are also highly valued for their medicinal properties. Most are cultivated, though *Ganoderma* spp. (Plate 9) are also collected from the wild. The

[3] Saprophyte describes a plant that feeds by external digestion of dead organic matter.

list of symbiotic macrofungi with medicinal properties is a short one, though there is some indication that they have been studied less because they cannot be cultivated (Reshetnikov, Wasser and Tan, 2001).

Symbiotic fungi

The most common form of symbiosis associated with wild edible fungi is that known as a mycorrhiza (Plate 2). Many plants depend on these fungus-root associations for healthy growth. A special type known as an ectomycorrhiza (ECM) is found on trees growing in the Taiga in the Russian Federation and the rain forest of Borneo and includes legume trees as well as conifers (Table 3). Ectomycorrhiza are typically formed by macrofungi and they include many of the key edible species that are collected in the wild, such as chanterelles (*Cantharellus* spp.) and *Amanita* species.

The mycorrhiza helps the tree to grow in nutrient-poor soils, such as the miombo woodland of central and southern Africa (Campbell, 1996). A sheath of hyphae wraps around the root. They penetrate the root structure but not the actual root cells themselves, forming a living contact between the fungus and the tree. The fungus helps the tree gather water from a wider catchment and delivers nutrients from the soil that the tree cannot access. The tree provides the fungus with essential carbohydrates.

Termitomyces contains important wild edible species. These fungi only grow in association with termites and their nests and are dependent on the organic matter brought by the insects from their feeding on trees. Although *Termitomyces* are saprobic, they are symbiotic with termites. Twenty edible species of *Termitomyces* have been recorded from Africa and Asia (Pegler and Vanhaecke, 1994). They are regularly collected and also sold (Plate 6). *T. titanicus* is the world's largest edible fungus, although other species are much smaller.

Rural people have long associated the appearance of edible fungi with particular trees and have incorporated this in local names. In southern Africa, *chimsuku* and *kamsuku* both describe *Lactarius* spp. that grow under *masuku* trees (Piearce, 1981). Some edible ectomycorrhizal fungi produce their fruiting bodies underground. The best known examples are the truffles (*Tuber* spp.: Plate 4). Over 400 species of edible ECM have been recorded (Wang, Buchanan and Hall, 2002). There are also many ectomycorrhizal fungi which produce fruiting bodies that are not edible or are poisonous.

The production of fruiting bodies depends on a complex set of factors and in some years production can be negligible. In Botswana, 14 tonnes of *Terfezia pfeilii*, one of the "desert truffles", were bought from one small community in one season; the next year only four fruiting bodies were located over a much larger area (Taylor, 2002, personal communication: *Edible fungi eaten and traded in Botswana and Namibia*). The lack of certainty of harvests from one year to the next makes it difficult to plan commercial exploitation and some attempts have been made to overcome this by "cultivating" key mycorrhizal species such as *Tricholoma matsutake* (Hall *et al*., 1998). Trees are successfully infected with truffles (Hall, Zambonelli and Primavera, 1998) and managed under controlled conditions in Italy (Plate 4) and elsewhere, but the time, effort and money required are only justified – assuming a good knowledge of the ecology of the fungus concerned – for the most valuable edible mycorrhizal species.

TABLE 3

Plant families with edible ectomycorrhizal fungi

FAMILY	EXAMPLES
Betulaceae	*Betula* (birches)
Caesalpinioideae	*Afzelia, Brachystegia, Isoberlinia, Julbernardia*
Casuarinaceae	*Casuarina*
Cupressaceae	*Cupressus*
Dipterocarpaceae	*Shorea, Dipterocarpus, Monotes*
Euphorbiaceae	*Uapaca*
Fagaceae	*Castanea* (chestnut), *Castanopsis, Fagus* (northern beech), *Nothofagus* (southern beech), *Quercus* (oak)
"Legumes"	*Acacia*
Myrtaceae	*Eucalyptus*
Pinaceae	*Pinus* (pines), *Picea* (spruces), *Abies* (firs), *Larix* (larches)
Papilionoideae	*Pericopsis*
Nyctaginaceae	*Neea*

For details of ectomycorrhizas on tropical trees, see Alexander and Hogberg (1986).

Tree species can form mycorrhizas with more than one fungus, and a fungus may associate with more than one tree. Some ECM are "native" to a region: in Madagascar an edible *Russula* grows on exotic eucalyptus (Buyck, 2001). Other edible ECM have been introduced and *Boletus edulis* is now found throughout southern Africa following the establishment of pine plantations. ECM have been most intensively studied in the past on temperate tree species but there have also been steady advances on tropical ECM in Africa (Thoen, 1993; Verbecken and Buyck, 2002).

Lichens are "self-supporting" associations between fungi and an alga or cyanobacterium and are the final example of a symbiosis that has edible properties. A lichen is a biological and not a systematic group (Kirk *et al.*, 2001) and several valuable species are eaten by people in Europe, Asia and North America and used for other economic purposes. They are not included in this book. Further information is available from a number of sources (e.g. Richardson, 1991; Marles *et al.*, 2000).

Plant pathogens and parasitic fungi

In several countries people eat plant material infected with plant pathogenic fungi. Maize cobs infected with the smut fungus *Ustilago maydis* are consumed in large quantities in Mexico, both fresh and canned. They are known locally as *huitlacoche* or *cuitlacoche* (Villanueva, 1997). *U. maydis* is a microfungus: it does not form a visible fruiting body and the only signs of its presence are a mass of dark spores (Plate 1). The cobs appear to become sweeter as the result of fungus attack (Sommer, 1995), and similar changes have been noted for the edible rust fungus *Cronartium conigenum* on pines in Mexico.

Other examples include: *Ustilago esculenta* on wild rice; *Sporisorium cruenta* on sorghum in China (Guozhong, 2002, personal communication: *Eating* Sporisorium cruenta *in China*); winged bean infected by *Synchytrium psophocarpi* in Indonesia (Rifai, 1989).

Hypomyces lactifluorum is a parasite macrofungus that grows on other macrofungi (boletes). It is eaten from Canada through to Guatemala and completes the range of ecological niches occupied by wild edible fungi.

IDENTIFICATION

Local and scientific names

Local names have been well documented in Mexico (Guzmán, 1997), China (Mao, 2000) and can be checked online for Malawi (www.malawifungi.org)[4] against the equivalent scientific names. Each of these countries has a rich lexicon of names and terms (Figure 1), a sign of the importance of wild edible fungi to rural people. Some local names have been adopted more widely, particularly for valuable edible fungi. *Boletus edulis* is commonly referred to by its French (*cèpe*) or Italian name (*porcino* – plural *porcini*), and *Tricholoma matsutake* by its Japanese name of *matsutake*.

The system of scientific names aims to remove doubt about the fungus being described. A person with *Cantharellus cibarius* in Nepal knows they have the same fungus as someone in Mozambique, assuming both have been accurately identified. The scientific name or binomial has two parts. The first name is the genus (*Cantharellus*) followed by the species name (*cibarius*). Named varieties exist for some species but their scientific validity is often uncertain.

Local names for edible fungi are based on shape, taste and other properties that are distinctive or important to people. The lichen (*Umbilicaria esculenta*) and an edible fungus (*Auricularia auricula-judae*) have similar common names in Hunan – *Yan-er* (ear of a rock) and *Mu-er* (ear of wood) respectively. This identifies where they grow

[4] All Web pages have been viewed in 2003.

and can be collected. Mycologists are sometimes wary of local classifications because they are based on scientifically unreliable characters (Härkönen, 2002).

Local names provide important clues to the uses and importance of edible fungi to people and there is much to be gained from their study. Local names allow researchers to learn about collecting practices, to analyse markets and to talk with forest managers and others who lack formal training in science and are unfamiliar with genera and species names. Examples of ethnographic studies involving wild useful fungi are listed in Table 13. Guidelines for conducting such studies are available from a number of different sources (e.g. Alexiades, 1996).

FIGURE 1
Naming the parts of a mushroom

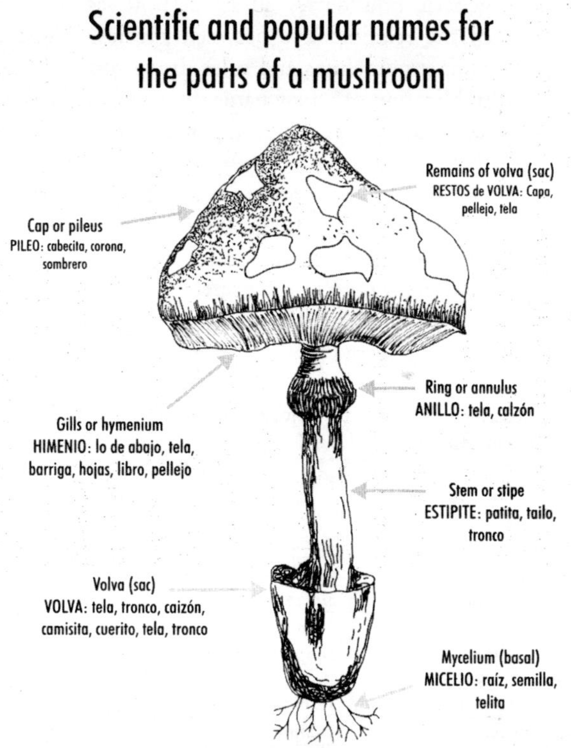

This example is based on a fruiting body of an *Amanita*. Other genera lack a volva (the sac that encloses the expanding fruiting body) and the ring may be absent. The English name is in bold; Spanish in capitals followed by popular names from Ajusco and Topilejo in Mexico.
Source: adapted from Reygadas, Zamoni-Martinez and Cifuentes, 1995.

Local and scientific classifications serve two different groups of people and neither is infallible. Edible species of *Boletus* are not eaten in parts of the United Republic of Tanzania, for example (Härkönen, 2002), reflecting local custom rather than scientific fact. Field guides often disagree on which species are edible, either because they are cautious about recommending species that require pre-cooking or because the authors are unaware of local customs in different parts of the world.

What is clear, however, is that there are many poorly described species sold and collected for personal use in developing countries. The rate of discovery is directly related to funding for projects and the ability to draw upon mycological expertise from different countries. Work in the United Republic of Tanzania (Härkönen, Codjia and Yorou, 1995), Mozambique and Malawi (Boa *et al.*, 2000), Burundi (Buyck, 1994b) and Benin (De Kesel, Saarimäki and Mwasumbi, 2002) emphasizes the richness of the tropical, edible mycota and how much remains to be done. In the absence of such mycological expertise local names can provide useful information, particularly if dried specimens are available for later examination.

An accurately identified specimen with a scientific name for that species ensures that any new knowledge can be reliably used. A scientific name is the most useful way of finding out whether a species is edible or poisonous, or if it has medicinal or other useful properties. An importer does not need to know if the *pied de mouton* from Bulgaria is *Hydnum repandum* since the genus contains only edible species, but an Italian buyer will pay less for the ordinary *Tuber sinosum* from China compared with other more valuable species. In this instance a scientific name reliably and uniquely describes the fungus in question, for which information can be gleaned from the literature.

Using the current or "correct" scientific name for a fungus

The scientific names for fungi are constantly changing – an indication of how much there is still to discover about the diversity of species. New names are proposed and generic boundaries adjusted, both as the result of new discoveries and a revision of the relationships between species. When a new species is proposed it is judged against guidelines and rules drawn up and regularly revised by scientists. The correct publication of a new name does not mean that scientists agree on its taxonomic status. The boundaries between genera and species are open to different interpretations and that is why there are "preferred" rather than "correct" scientific names for fungi.

These changes and uncertainties have important practical consequences for people using wild edible fungi. People have to be aware that a species was previously known by a different name or synonym when searching for information: *Termitomyces albuminosus* was once known as *Collybia albuminosa.* Other changes are less dramatic. *Lentinus edodes,* or *shi'itake* now has the preferred name of *Lentinula edodes.* The older "non-preferred" name is still regularly used in publications. Opinions are still divided as to whether *Coriolus* species with medicinal properties should be renamed *Trametes. Auricularia auricula-judae,* the "preferred name", appears variously as *Hirneola auricula-judae* and *Auricularia auricula*.

Table 4 lists the preferred names of wild edible fungi that are still commonly referred to by other names. Common spelling mistakes also appear in publications; even minor differences can cast doubt on the identity of a fungus. The *Dictionary of the fungi* is a standard reference that is regularly revised to list all genera of fungi (Kirk *et al.* 2001). Index Fungorum, an Internet resource, allows users to check the preferred or non-preferred status for species names and to find synonyms (www.indexfungorum.org). This is of considerable practical benefit, although Index Fungorum lacks the backing required to answer fully queries about which scientific name to use for wild edible fungi. This practical need has still to be addressed by the scientific community.

TABLE 4
Preferred (current or "correct") names of economically important wild fungi

As published	Preferred name
Armillariella mellea	*Armillaria mellea*
Auricularia auricula	*Auricularia auricula-judae*
Xerocomus badius	*Boletus badius*
Boletus granulatus	*Suillus granulatus*
Boletus luteus	*Suillus luteus*
Calvatia gigantea, Lycoperdon gigantea	*Langermannia gigantea*
Collybia albuminosa	*Termitomyces albuminosus*
Coriolus hirsutus	*Trametes hirsuta*
Coriolus versicolor	*Trametes versicolor*
Dendropolyporus umbellatus	*Polyporus umbellatus*
Fomitopsis officinalis	*Laricifomes officinalis*
Grifola umbellatus	*Polyporus umbellatus*
Hericium erinaceum +	*Hericium erinaceus*
Hirneola auricula-judae	*Auricularia auricula-judae*
Hydnum imbricatus	*Sarcodon imbricatus*
Hypsizygus ulmarium	*Lyophyllum ulmarium*
Lentinus edodes	*Lentinula edodes*
Lepiota procera	*Macrolepiota procera*
Lepiota rhacodes	*Macrolepiota rhacodes*
Panus rudis	*Lentinus strigosus*
Pleurotus cornucopiae var. citrinopileatus	*Pleurotus citrinopileatus*
Pleurotus ferulae	*Pleurotus eryngii var. ferulae*
Pleurotus olearius	*Omphalotus olearius*
Pleurotus opuntiae	*Pleurotus ostreatus*
Pleurotus porrigens	*Pleurocybella porrigens*
Pleurotus tuber-regium	*Lentinus tuber-regium*
Poria cocos; Wolfiporia cocos	*Wolfiporia extensa*
Rozites caperata +	*Rozites caperatus*
Sparassis radicata	*Sparassis crispa*
Strobilomyces costatispora	*Afroboletus costatisporus*
Termitomyces eurrhizus +	*Termitomyces eurhizus*
Tricholoma gambosa	*Calocybe gambosa*
Tricholoma lobayensis; T. lobayense	*Macrocybe lobayensis*
Verpa bohemica	*Ptychoverpa bohemica*

See www.indexfungorum.org for further advice and information.
+ indicates a common misspelling.

Identifying species

The genera of wild edible fungi found in tropical and subtropical climates are broadly similar to those found in the mycota of temperate regions (Lincoff, 2002). The species diversity is, however, much greater in developing countries and care must be taken when comparing specimens with the narrower range of species illustrated in the many field guides published in Europe and North America.

Edible fungi occur in two major taxonomic groups. The basidiomycetes contain the mushrooms, bracket fungi and boletes (Plate 1); the ascomycetes include truffles (Plate 4) and morels (Plate 9). There is no simple test for determining edibility. The scientific literature is the best objective source of advice, but local practices and preferences can also reveal useful information. Empirical evidence is the ultimate indication of whether or not a species is edible.

The classical method for identifying a macrofungus involves a microscopic examination of tissues, spores and sporing structures. This will at least ensure that the genus is identified. Identification of the lesser known tropical species may also require examination of reference collections (Plate 3). Useful visual clues can be obtained from photographs in field guides and there are increasing numbers of Web sites with photographs and written descriptions of species (Chapter 6). Information on Mexican NWFP provided by the Secretaría de Medio Ambiente y Recursos Naturales (2002) on the Internet includes wild edible fungi and is an excellent example of an online guide that could be developed for other regions (see www.semarnat.gob.mx).

Expert identifications can be costly, although paying for an identification does provide a guarantee of getting a response to a query (Meijer, 2001). Preserving specimens is always useful and at its simplest provides a local reference for comparing specimens. Most macrofungi are easily preserved by drying (Halling, 1996). There are special drying racks for fungi (Plate 3), but these can also be locally improvised, adopting methods used for drying fruits and other food produce. Dried specimens can, if necessary, be sent at a later date for scientific identification and should be accompanied by field notes and/or colour photographs.

Molecular tools are commonly used to identify plant pathogenic fungi and have also been applied to truffle species in order to detect which species are used in prepared foods. The practical application of these tools for identifying and characterizing edible macrofungi has still to be explored.

Sources of technical advice and support are discussed in Chapter 6.

MAJOR GROUPS OF WILD FUNGI

There are more than 200 genera of macrofungi which contain species of use to people, mostly because of their edible properties. A clear distinction is made in this book between those recorded as simply "edible" and those that are actually eaten ("food"). To include all edible species as "food" would greatly overstate the number of species consumed by people around the world. Wild fungi with medicinal properties are also valued by rural people in several countries, though this is of secondary importance.

The major genera of wild edible fungi are described in Table 5, with brief notes on medicinal species. The genera of wild edible fungi can be divided into two categories: those containing species that are widely consumed and often exported in significant quantities, such as *Boletus* and *Cantharellus*; and those with species that are eaten widely, usually in small amounts, and rarely if ever traded beyond national boundaries. Annex 1 summarizes the general importance of wild edible fungi by country while Annexes 2 and 4 list individual species.

Medicinal mushrooms

Medicinal mushrooms are attracting greater scientific and commercial interest, prompted by a renewed awareness of the use of such material in traditional Chinese medicine (Table 17). The *International Journal of Medicinal Mushrooms* began publication in 1999 and is an important source of information for this expanding field of research (Wasser and Weis, 1999b). See Chapter 4 for further discussions about the health benefits of medicinal mushrooms.

Ceremonial aspects

The ceremonial and religious roles played by wild fungi in different cultures are closely associated with hallucinogenic properties. This has attracted much scientific

TABLE 5
Important genera of wild fungi with notes on uses and trade

Information obtained mostly from developing countries. See www.wildusefulfungi.org for more details of individual records for species and countries. "Food" signifies confirmed use of species; "edible" is a noted property *without* confirmed consumption. The total number of edible species is the sum of the two. Use refers to country of origin and not countries of export. "Medicinal" ('med.') is a noted property and does not confirm use of species for health reasons. Edible species may have medicinal properties and therefore the total number of species in bold may be less than the sum of individual uses. See Lincoff (2002) for distribution of major groups of edible fungi around the world.

Genus	No. of species use and properties	Country use and general notes
Agaricus	**60**	Edible species reported from 29 countries, as food in 13 (under-reported, though note possible confusion between wild and cultivated sources).
	food 43 *edible* 17 *med.* 6	Agaricus species are regularly collected from the wild but only cultivated forms are exported. Some species are poisonous. *A. bisporus* is the mostly commonly cultivated edible fungus. The medicinal *A. blazei* is exported from Brazil to Japan and cultivated and sold in China.
Amanita	**83**	Edible species reported from 31 countries; as food in 15 (under-reported).
	food 42 *edible* 39 *med.* 7	*A. caesarea* is highly valued in countries such as Mexico, Turkey and Nepal. Few species are traded across national borders. There are a notable number of poisonous species. *A. phalloides* is a major cause of deaths around the world from consumption of wild fungi.
Auricularia	**13**	Edible species reported from 24 countries, as food in 10 (under-reported).
	food 10 *edible* 3 *med.* 4	A global genus with a relatively small number of species. Known generically as "ear fungi", they are distinctive, easily recognized and consumed by forest dwellers in Kalimantan as well as rural communities in all continents. Some species have medicinal properties. There is a major trade in cultivated species though few data have been seen. Key species: *A. auricula-judae*
Boletus	**72**	Edible species reported from 30 countries; as food in 15 (under-reported)
	food 39 *edible* 33 *med.* 7	*B. edulis* is the best known species, regularly collected and sold and major exports from outside and within Europe. There are a some poisonous species but few incidents. "Bolete" is a general description of a macrofungus with a stalk and pores on the underside of the cap. Apprehension exists about eating "boletes" in east and southern Africa.
Cantharellus	**42**	Edible species reported from 45 countries; as food in 22 (under-reported).
	food 22 *edible* 20 *med.* 3	A diverse and cosmopolitan genus containing widespread species such as *C. cibarius*. Sold in markets in many countries, sometimes in functional mixtures of different species. Major quantities are collected and exported around the world. No poisonous species.
Cordyceps	**37**	Useful species (mostly medicinal) reported from three countries.
	edible? 35 *med.* 9	The only reason for 'eating' species is for health benefits. Collected intensively in parts of China and less so in Nepal. Many species described from Japan, but local use uncertain. Widely valued for its medicinal properties and an important source of income for collectors. Key species: probably *C. sinensis* and *C. militaris*
Cortinarius	**50**	Edible species reported from 11 countries; as food in three.
	food 30 *edible* 20 med. 10	Widely disregarded in Europe and North America because of concern about poisonous species. Most records of local use are restricted to a few countries e.g. China, Japan, the Russian Federation and Ukraine. No known export trade.
Laccaria	14	Edible species reported from 17 countries; as food in four (under reported)
	food 9 *edible* 5 *med.* 4	Regularly collected and eaten, also sold widely in markets. No reports of export trade, which is unsurprising given their generally small size and unremarkable taste. Key species is *L. laccata*.
Lactarius	**94**	Edible species reported from 39 countries; as food in 17 (under reported).
	food 56 *edible* 38 *med.* 7	Many different species are regularly collected and eaten. Key species such as *L. deliciosus* are highly esteemed and there is a valuable trade in Europe. Several key species frequently sold in local markets. Little reported export activity despite widespread popularity, perhaps reflecting the diversity of species on offer.
Leccinum	**22**	Edible species reported from eight countries; as food in two.
	food 4 *edible* 9	Widely eaten and collected but little trade beyond national boundaries. Key species *L. scabrum*. Possible exports from pine plantations in tropics, but poorly understood.
Lentinula	**3**	Edible species reported from six countries; as food in four.
	food 2 *edible* 1 *med.*1	*Lentinula edodes* is the key species (= *Lentinus edodes*). Known as *shi'itake* it is cultivated in many countries and is an important commercial species (nearing 30% cultivated amount). Cultivated *shi'itake* is exported.
Lentinus	**28**	Edible species reported from 24 countries; as food in eight (under-reported).
	food 16 *edible* 12 *med.* 5	Although many different species are collected and used locally only two or three are of any significance. Key species probably *L. tuber-regium*, valued for its medicinal properties. Little or no export trade.

Genus	No. of Species use and properties	Country use and general notes
Lycoperdon	**22**	Edible species reported from 19 countries; as food in seven (under-reported).
	food 9 *edible* 10 *med.* 10	There are many records of species being eaten but typically reports are of small-scale collecting and use. Only market sales known are in Mexico. Key species are *L. pyriforme* and *L. perlatum*.
Macrolepiota	**13**	Edible species reported from 33 countries; as food in nine (under-reported).
	food 7 *edible* 6 *med.* 1	M. procera is the key species and most recorded, from around 15 countries on all major continents. Locally consumed; trade is essentially small-scale and local.
Morchella	**18**	Edible species reported from 28 countries; as food in 10 (under recorded).
	food 14 *edible* 4 *med.* 5	Highly valued genus with several species that fruit in abundance in certain years and are a major source of (export) revenue in several countries. Species are not always eaten in countries where they are collected. Key species *M. esculenta*.
Pleurotus	**40**	Edible species reported from 35 countries; as food in 19 (under reported).
	food 22 *edible* 18 *med.* 7	Key species is *P. ostreatus* in terms of amounts eaten, predominantly from cultivation. Other species said to be more tasty. Species occur widely and are regularly picked though seldom traded from the wild.
Polyporus	**30**	Edible and medicinal species reported from 20 countries; as food or medicine in seven.
	food 15 *edible* 9 *med.* 12	Many species are regularly used and eaten but of relatively minor importance. Some are cultivated. Only one record known, from Nepal, of selling in markets. No international trade is known to occur.
Ramaria	**44**	Edible species reported from 18 countries; used as food in seven.
	food 33 *edible* 11 *med.* 5	Many records of local use. Regularly sold in markets in Nepal and Mexico and elsewhere. Several major species but perhaps *R. botrytis* is the most commonly collected and used. Some species are poisonous, others are reported to have medicinal properties.
Russula	**128**	Edible species reported from 28 countries; as food in 12 (under-reported).
	food 71 *edible* 54 *med.* 25	One of the most widespread and commonly eaten genera containing many edible species. Also poisonous varieties though most can be eaten after cooking. Regularly sold in markets but species names not always recorded. Genus is of tropical origin. Notable species include *R. delica* and *R. virescens*.
Suillus	**27**	Edible species reported from 25 countries; as food in 10 (under-recorded).
	food 26 *edible* 1 *med.* 2	Key species is *S. luteus*, exported from Chile. *S. granulatus* is more widely recorded though its use as a food is limited. Many other species are regularly collected and eaten and several are sold in Mexican markets.
Terfezia	**7**	Edible species reported from eight countries; as food in four.
	food 5 *edible* 2	Desert truffles occur widely in North Africa and parts of Asia. They are said to be important but few details were found concerning trade or market sales.
Termitomyces	**27**	Edible species reported from 35 countries; as food in 16 (under-reported).
	food 23 *edible* 4 *med.* 3	Highly esteemed genus. Many species are widely eaten with often high nutritional value. Collected notably throughout Africa. Used widely in Asia but less well documented. Notable species include *T. clypeatus, T. microporus* and *T. striatus*. Sold in markets and along roadsides, and good source of income.
Tricholoma	**52**	Edible species reported from 30 countries; as food in 11 (under-reported).
	food 39 *edible* 13 *med.* 17	The most important species is *T. matsutake*, in terms of volume collected and financial value. China, both Koreas and the Russian Federation are major exporters to Japan. The Pacific northwest of North America, Morocco and Mexico export related species, but only in significant quantities from the first. Some species are poisonous if eaten raw; others remain so even after cooking. Ignored or lowly esteemed in several countries prior to export opportunities e.g. Bhutan, Mexico (Oaxaca).
Tuber (truffles)	**18**	Edible species reported from eight countries; as food in four (under-reported).
	food 8 *edible* 10	Contains species of extremely high value and much esteemed in gourmet cooking, but only of very minor significance to poor communities in the South. There is some interest from Turkey in management of truffles. Scientific principles have been applied to truffle management and successful schemes initiated in Italy, France, Spain and New Zealand. The "false truffles" comprise other genera e.g. *Tirmania, Rhizopogon, Terfezia*.
Volvariella	**12**	Edible species reported from 27 countries; as food in 7 (under-reported, though note possible confusion between wild and cultivated origins).
	food 5 *edible* 7 *med.* 1	Key species is *V. volvacea*. Widely cultivated and sold in local markets but also collected from the wild.

TABLE 6
Fungi with conflicting reports on edibility

Binomial	Notes*
Agaricus arvensis	Reported mostly as edible and eaten in Mexico; also said to be a gastrointestinal irritant (Lincoff and Mitchel, 1977).
Agaricus semotus	Said to be edible from Hong Kong (Chang and Mao, 1995); others say it is poisonous (Rammeloo and Walleyn, 1993).
Amanita spissa	Several reports indicate this can be eaten (although none state "food"); an equal number say it is poisonous, e.g. Chang and Mao, 1995.
Amanita flavoconia	Conflicting accounts from Mexico: one report says it is edible, the other that it is poisonous.
Amanita gemmata	Reported as edible from Mexico and Costa Rica but implicated in a poisoning case from Guatemala (Logemann *et al.*, 1987).
Boletus calopus	Edible in the Russian far east (Vasil'eva, 1978); said to be poisonous in Slovenia (www.matkurja.com) and by other field guides.
Chlorophyllum molybdites	Many reports confirm that this is a poisonous species but it is also said to be edible in Mexico (Villarreal and Perez-Moreno, 1989) and Benin (De Kesel, Codjia and Yorou, 2002). Easily confused with *Macrolepiota procera*, a well known edible species.
Coprinus africanus	Eaten in Nigeria (Oso, 1975); other reports suggest it is poisonous in Africa (Walleyn and Rammeloo, 1994).
Coprinus atramentarius	Edible if eaten in the absence of alcohol; this produces an unpleasant effect if imbibed at the same time, hence remarks that it is potentially poisonous (Lincoff and Mitchel, 1977).
Gyromitra esculenta	In Finland it is a delicacy (Härkönen, 1998) and it is also widely eaten in the Russian Federation and neighbouring regions. In other countries it is said to be poisonous and can kill when raw (Hall *et al.*, 1998a). The toxic properties are mitigated by suitable preparation prior to eating.
Gyromitra infula	Eaten in Mexico (www.semarnat.gob.mx) but also reported as poisonous (Lincoff and Mitchel, 1977).
Helvella lacunosa	Widely eaten but also reported as toxic if eaten raw (Lincoff and Mitchel, 1977).
Lactarius piperatus	Many reports say it is edible and confirmed as food in Turkey (Caglarirmak, Unal and Otles, 2002) but also reported as poisonous in China (Liu and Yang, 1982).
Lactarius torminosus	Several reports say it is edible (e.g. Malyi, 1987); others say it is poisonous (Hall *et al.*, 1998a).
Lampteromyces japonicus	A common cause of poisoning in Japan (Hall *et al.*, 1998a) but also has medicinal properties (Hobbs, 1995).
Lenzites elegans	Edible in the United Republic of Tanzania (Rammeloo and Walleyn, 1993) but maybe poisonous in the Democratic Republic of the Congo (Walleyn and Rammeloo, 1994).
Lepiota clypeolaria	Edible in Mexico and Hong Kong Special Administrative Region, China, but also said to be poisonous.
Morchella esculenta	Like other morels said to be poisonous if eaten raw (Lincoff and Mitchel, 1977). Edible and good when cooked.
Paxillus involutus	Widely reported as poisonous but said to be edible after suitable cooking and preparation in the Russian far east (Vasil'eva, 1978).
Phallus indusiatus	Reported as edible (Bouriquet, 1970) and poisonous (Walleyn and Rammeloo, 1994): both reports are from Madagascar.
Podaxis pistillaris	Reported as edible from India and Pakistan (Batra, 1983). Said to be poisonous in Nigeria (Walleyn and Rammeloo, 1994); medicinal properties (Hobbs, 1995).
Ramaria formosa	Edible in Nepal (Adhikari and Durrieu, 1996) but said to be poisonous in several other countries, including Bulgaria (Iordanov, Vanev and Fakirova, 1978).
Russula emetica	Undoubtedly poisonous if eaten raw but said to be edible in Mexico (Zamora-Martinez, Alvardo and Dominuez, 2000) and the Russian far east (Vasil'eva, 1978).
Stropharia coronilla	Conflicting reports within Mexico: said to be edible (Villarreal and Perez-Moreno, 1989) and poisonous (Aroche *et al.*, 1984).
Suillus placidus	Said to be edible (Vasil'eva, 1978) and poisonous (Chang and Mao, 1995).
Tricholoma pessundatum	Edible in Hong Kong (Chang and Mao, 1995) but *T. pessundatum* var. *montanum* reported as poisonous elsewhere (Lincoff and Mitchel, 1977).
Tricholoma sulphureum	All records say it is poisonous apart from an account from India that says it is edible (Purkayastha and Chandra, 1985).

and personal interest, particularly in Mexico (Davis, 1996; Riedlinger, 1990). Globally this use of wild fungi is of minor or no relevance to most countries.

EDIBILITY AND POISONOUS FUNGI

Many macrofungi are not worth eating or are simply inedible. This worthless group of species – as defined by their edibility – significantly dwarfs the very small number of toxic or poisonous species, of which there are only a very few that can kill. Yet it is also true that this very small group of lethal species has significantly shaped attitudes to eating wild fungi, creating potential barriers to wider marketing in many places.

Knowing the scientific name of a fungus provides a good indication of its edibility. In some cases the genus alone will suffice; all known *Cantharellus* species are edible (though not equally tasty). On the other hand, *Amanita* contains both exquisite edible and deadly poisonous species. The only reliable guide to edibility is the knowledge that someone has eaten a particular type – and survived. Local practices and preferences are therefore another useful source of information.

There are conflicting reports in field guides about edibility. Some recommend eating species that others reject as poisonous. People from eastern Finland regard the false morel, *Gyromitra esculenta,* as a culinary delicacy once it has been carefully pre-cooked. Guides in the United States and elsewhere state emphatically that the fungus is poisonous and should not be eaten. Other examples of conflicting advice are summarized in Table 6.

What species are eaten?

> *Reports of edible and poisonous species are based on named sources. The accuracy of this information lies with these original sources.*

A total of 1 154 edible and food species have been recorded from 85 countries (Table 1). The species eaten in one country or region often differ from nearby areas and in some cases there are dramatic changes in tradition. The Mesoamerican tradition of eating wild edible fungi continues from Mexico to west Guatemala then is absent from much of Honduras and Nicaragua, even though both contain forest areas that in theory support production of edible fungi.

The number of species eaten is sometimes only a fraction of those available. Only 15 of the 284 edible species in Armenia are regularly eaten (Nanaguylan, 2002, personal communication: *Edible fungi in Armenia*). In two districts of Turkey, 12 out of a possible 29 edible species were collected and eaten (Yilmaz, Oder and Isiloglu, 1997). The reasons for these different patterns of use are not always clear but there is a trend of less frequent use as people move away from the land. Rural people in Guatemala have a positive yet informed approach to eating wild fungi which people living in cities lack (Lowy, 1974). Educated people living in towns in Malawi lose the strong local traditions that rural communities maintain and even acquire a suspicious attitude towards wild fungi (Lowore and Boa, 2001).

In parts of the United Republic of Tanzania boletes are thought to be poisonous (Härkönen, Saarimäki and Mwasumbi, 1994a). In Colombia there is no apparent tradition of eating wild fungi in the Andean regions, though they occur widely (Franco-Molano, Aldana-Gomez and Halling, 2000). *Tricholoma matsutake* was of little local interest in Sichuan, China (Winkler, 2002) prior to Japanese demand that stimulated an export trade in the late 1980s and appears to have prompted wider local consumption. A similar event took place in the Pacific northwest, though with *Tricholoma magnivelare* (Redhead, 1997). This was collected and eaten by Japanese settlers in the 1930s (Zeller and Togashi, 1934) but at the time this did not arouse much, if any, local interest.

Poisonous species

A review of poisoning incidents in official and informal publications shows that the frequency of such events and the effect on humans are overall less than that suggested by attendant publicity (Logemann *et al.*, 1987). During the search for information on wild edible fungi, about 170 poisonous species were noted. Most are either related to edible species or confused with them. There are, of course, real dangers in collecting and consuming poisonous fungi, but these should be seen against the wider background of millions of people collecting and eating wild fungi safely on a regular basis.

Several popular and highly esteemed edible species are poisonous when raw. Few people eat them in this condition and risks of poisoning are in reality small. Poisonous mushrooms vary in their effects from mild stomach and digestive upsets to more serious problems such as liver damage. The solutions to these potential risks include providing local advice on which species to collect and which ones to avoid (Plate 3) and publicity campaigns that highlight potentially poisonous species on posters. Mr Sabiti Fides, a trader in Malawi, took a more direct route by eating mushrooms in front of his customers (Box 3).

In southern Africa roadside sellers only offer "safe species" (Ryvarden, Piearce and Masuka, 1994) and most market places are a reliable means of obtaining known, edible wild fungi. Problems can occur with "contamination" in markets but such incidents are most uncommon (see Table 8).

Finland has trained mushroom advisers covering all rural areas (Härkönen, 1998; Härkönen and Järvinen, 1993). The *svamp* "police" based in some town centres in Norway help collectors identify edible species, and there are similar schemes in other countries.

Poisonings are associated with a number of events:

- young children collecting indiscriminately and eating raw mushrooms;
- immigrants arriving in a new country and wrongly identifying a local species that turns out to be poisonous;
- food shortages and economic hardship force people to hunt for food;
- different physiological responses to an "edible" fungus.

Mexicans living in California have eaten *Amanita phalloides* – a poisonous species not found at home – thinking it was the edible *Volvariella volvacea* (Plate 2). The guide for edible mushrooms in Israel is written in Hebrew and Russian (Wasser, 1995), following the arrival of over one million Russians in the 1990s and their strong tradition of collecting wild edible fungi. One Russian was poisoned when he too

BOX 3

"If I eat this *bowa* it is OK to buy" – Mr Sabiti Fides, trader from Malawi

"We asked around for a typical *bowa** 'middleman' or 'wholesaler' and met with Sabiti Fides. As it turned out he was not typical at all but really rather exceptional – the KING OF THE *BOWA* TRADERS. Fides started buying *bowa* from Machinga and taking them to Zomba for sale in the 1998-99 season. He was trying to think of ways of earning some money to support his family. He observed that at the end of a day on the roadside stall a good deal of *bowa* remained unsold. He decided to buy them up and take them to Zomba.

In order to find customers he would walk around residential areas such as the police training college, the barracks, Chancellor College and also the suburbs such as Mponda Bwino and Chikanda, selling from house to house. At first he found the householders reluctant – 'maybe they are poisonous', 'maybe they are not good'. Patiently he would persuade the buyers (mainly women) to try them – tasting some himself in order to demonstrate lack of poison. One might buy. Then the next time others would have observed that the one who bought enjoyed their purchase and they would follow suit. Gradually he would build up his regular customers who eventually would buy without fail."

* *bowa* – edible fungus

Source: Lowore and Boa (2001).

TABLE 7
Incidents of large-scale poisoning caused by consumption of wild fungi

CHINA	Number dead	Number poisoned	Notes
1962–82	108	444	Ninghua county, Fujiang province (Liu and Yang, 1982): 88 incidents were reported. Of the 16 poisonous species known to occur, 11 belong to *Russula* or *Amanita*. Population of Fujiang in 2000 was 34 million.
2001	6	1 700	People bought "poisonous mushrooms" from a market. Report by Yongkiu county health bureau; via www.hclinfinet.com.
Total	**113**	**2 037**	
POLAND	**Number dead**	**Number poisoned**	**Notes**
1931	31	ns	All children and associated mainly with eating *Amanita phalloides.* Occurred in Poznan (Lincoff and Mitchel, 1977) – from an account by Simons (1971).
1952	11	91	Consumption of *Cortinarius orellanus* (Lampe and Ammirati, 1990).
1953–62	64	708	From a survey of incidents over a ten-year period. Further deaths and poisonings occurred from eating *Cortinarius orellanus, Gyromitra esculenta* (dead – 6; poisoned – 132) and principally *Amanita phalloides* (dead – 54; poisoned – 553). Lincoff and Mitchel, (1977) based on Grzymala (1965).
Total	**106**	**799**	
RUSSIAN FEDERATION	**Number dead**	**Number poisoned**	**Notes**
1992	23	170	Report in the *Los Angeles Times*, 8 August 1992. Occurred about 350 miles from Moscow. Species of fungi involved not mentioned.
1999	ns	2 240	From *Pravda*, 30 May, 2001. This short report says that the incidents occurred mostly in Central Russia.
2000a	ns	2 470	Also from *Pravda*, 30 May 2001, and again notes that the incidents occurred mostly in Central Russia.
2000b	ca. 30	ca. 300	Report from the *Los Angeles Times*, 16 July 2001, says that an "unusually high number of deaths" were reported by the local authorities in Belgorod, Voronezh and Volgogad Oblasts. They were linked to consumption of *Amanita phalloides* but other species may have been involved. Police patrolled forests to discourage collection and checked baskets of collectors.
Total	**53**	**5 180**	
UKRAINE	**Number dead**	**Number poisoned**	**Notes**
1992	40	400	Report from the *Los Angeles Times*, 8 August 1992. Species responsible for these incidents were not mentioned.
1998	74	ns	Associated Press, date unknown (www.geocities.com/Yosemite/Trails/7331).
1999	42	ns	As above.
2000	112	ns	As above.
Total	**268**	**400 (4 000*)**	

ns – not stated.
* Sum calculated using an estimated ratio of ten poisoned to each person who dies, to account for those years where people died but the number of people poisoned and who recovered were not stated.

mistook a poisonous species for an edible species known from his home country (Hazani, Taitelman and Sasha, 1983). Other reports suggest a certain recklessness amongst Russians in choosing which species to collect and eat (Matsuk, 2000).

Some people eat *Laetiporus sulphureus* without any ill-effects while others feel ill. The suggested reason is that physiological responses by people differ but there could also be different strains of the fungus, which differ in chemical composition. Little is known about this particular feature for poisonous or potentially poisonous species.

A summary of well-publicized incidents o widespread poisoning is given in Table 7. There has been a spectacular rise in poisonings and deaths in Ukraine in the last decade. Various reasons have been given, including a dramatic economic downturn and the desperate search for food[5] or produce to trade in local markets.

[5] "I had never seen people (in central Lviv) not only rummaging in dustbins, but putting valuable scraps of food from them directly into their mouth – even in the collapsed societies such as Georgia and Moldova." (Almond, 2002).

Regular reports of poisonings in the United States appear in the journal *McIlvainea* (e.g. Cochran, 1987). These incidents are insignificant by comparison with the thousands of people who collect and consume wild fungi without any reported problems. Millions of other people around the world also regularly eat wild edible fungi without any ill-health effects, and it is important to keep a sense of perspective when reviewing the reported incidents of poisoning.

Contamination of wild edible fungi

The Chernobyl accident in Ukraine in the 1980s prompted investigations of radioactive materials in sources of wild food and particularly wild edible fungi. Broader concerns about the accumulation of heavy metals and pollutants by macrofungi have also been expressed.

A study of radiocaesium intake via consumption of wild fungi in the United Kingdom concluded that intake depended more on the species eaten than the weight consumed (Barnett *et al.*, 2001). Mycorrhizal fungi had a significantly greater radioactivity compared to saprobic or parasitic species. Consumption of wild edible fungi in the United Kingdom is small by comparison with other countries but the study gives a general indication of the potential health risks.

One reported case of contamination concerned the accidental mixing of potentially poisonous wild species with wild edible fungi imported by the United States (Gecan and Cichowicz, 1993). Such events are rare, however, and there are no known instances of this causing any damage to human health in Europe.

CULTIVATION OF EDIBLE FUNGI

There are nearly a hundred species of fungi that can be cultivated (Annex 4). All are saprobic. Commercial markets are dominated by *Agaricus bisporus*, *Lentinula edodes* and *Pleurotus* spp. (Table 18) and these account for nearly three quarters of the cultivated mushrooms grown around the world (Chang, 1999). The major cultivated species are grown on a variety of organic substrates, including waste from producing cotton and coffee. The technologies are well established and successful mushroom industries have been established in many countries. There has been a huge increase in production in the last ten years, mostly as a result of increased capacity in China.

Reports from Africa (Mshigeni and Chang, 2000), Mexico (Martínez-Carrera *et al.*, 2001) and Amazonia in Brazil (Pauli, 1999) suggest that mushroom cultivation offers economic opportunities as well as nutritional and health benefits. Small-scale cultivation takes place throughout China and could provide a suitable model for technology transfer. The cultivation of the paddy straw fungus (*Volvariella volvacea*) is integrated with rice production in Viet Nam. Wherever saprobic species are cultivated they require a steady supply of raw materials. The expansion of *shi'itake* production in Qingyuan, China ("the mushroom capital of the world") led to a serious depletion of local forests that supplied the wood on which to grow this edible fungus (Pauli, 1998).

The number of saprobic species being cultivated is steadily increasing and information and practical advice are readily available (Stamets, 2000). Ectomycorrhizal fungi can also be "cultivated". Trees are inoculated with truffle fungus that must then infect the roots and form the ectomycorrhizae. The trees are carefully tended to encourage production of the truffles (Plate 4). Methods for "cultivating" truffles are constantly being refined and improved (Hall *et al.*, 1998a).

PLATE 1
TYPES OF MACROFUNGI

Edible fungi come in many shapes and sizes. There are no consistent features (or tests) that distinguish them from poisonous varieties. Examples are from Malawi and photos by Eric Boa, unless stated otherwise.

1.1 *Lactarius* sp. White fluid appears after breaking the gills. Many species are edible and all are mycorrhizal.

1.2 *Amanita loosii*, edible. The sac is a distinctive feature of *Amanita*, a genus that includes poisonous species. (photo: *Paul Kirk*)

1.3 Common ear fungus, *Auricularia auricula-judae*. Edible. France. Also widely cultivated.

1.4 *Ramaria* sp. There are a number of similar varieties eaten around the world

1.5 This *Afroboletus* has a dense network of tiny pores on the underside of the cap.

1.6 (left) *Lycoperdon* sp., Norway. Puffballs are widespread and eaten regularly, though in relatively small quantities.

1.7 (right) *Cantharellus* sp. The gills continue along part of the stem and the fruiting bodies have a distinctive appearance.

PLATE 2
HOW FUNGI GROW: mycorrhizas, saprobes and pathogens

Fungi obtain their food symbiotically, as saprobes or parasites (pathogens). There are edible macrofungi in each category. The most valuable wild species are ectomycorrhizal, a form of symbiosis. Ectomycorrhizal roots have a distinct though varied appearance. It is unusual to see them clearly *in situ*. Many saprobic macrofungi are edible. Few pathogens are eaten. All examples are from Malawi unless stated otherwise. All photos by Eric Boa.

2.1 Ectomycorrhiza. The white covering on the roots indicates the fungal sheath

2.2 This very distinctive yellow ectomycorrhiza is associated with a *Cantharellus* sp.

2.3 These ectomycorrhizas are small and fluffy. Mycelium in the soil can have a similar appearance.

2.4 Tracing a fungus back to the host tree is possible when a physical connection to the roots can be seen.

2.5 *Agrocybe aegerita*, an edible saprobic species growing here on a tree stump in Bologna, Italy. Also cultivated.

2.6 Paddy straw or *Volvariella volvacea*. Commonly cultivated, it is a saprobic fungus. Indonesia. Edible.

2.7 Maize cob infected by *Ustilago maydis*, Bolivia. Earlier stage infections are eaten as *huitlacoche* in Mexico.

2.8 *Armillaria mellea*, a tree pathogen, at the base of a dead laburnum tree. London. Edible

PLATE 3
WHICH FUNGI ARE EDIBLE? IDENTIFYING SPECIES

Edible species can be identified using local and scientific knowledge. Neither system is infallible: local practices are based on empirical evidence of edibility, though local beliefs may falsely exclude edible species. A scientific name provides access to published information on properties, but conflicting advice may exist. Used together, local and scientific knowledge are a powerful guide to properties of wild fungi. All photos by Eric Boa unless stated.

3.1 (left) This French pharmacy offers local assistance in identifying edible species

3.2 (right) The second oldest publication on wild edible fungi from China. It includes descriptions of 'species' and would have been a useful reference book. (photo: *Warren Priest*)

3.3 (left) Paul Kirk documents a field collection from Malawi. Each specimen is given a reference number and described before being dried, and thus preserved for further examination.

3.4 (right) Spore print of *Hypholoma fasciculare*, a poisonous species. The upper print is after leaving the cap for several hours; the one below for less than an hour. Spore colour helps to distinguish similar genera but not to species.

3.5 (right) Alessandra Zambonelli of the University of Bologna with a unique collection of truffle specimens from around the world. Collections are vital reference sources for identifying fungi and naming new species.

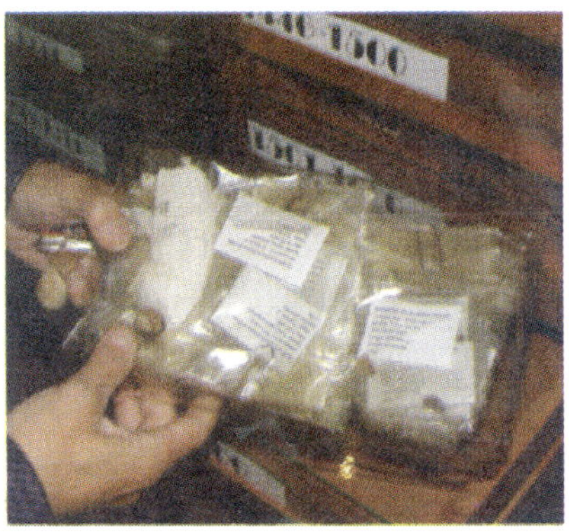

3.6 (left) Dried examples of truffles are carefully labelled and stored in the collection.

3 Management: wild edible fungi, trees, forest users

MULTIPLE USE OF FORESTS: ISSUES AND CONFLICTS

The management of wild edible fungi and their sustainable production must address two key topics: first, forests and their management and second, forest users. Successful management of wild edible fungi balances the impact and effects of collection and harvesting against the wider aims of forest management. These wider aims are determined by the relative importance of different forest uses. Are wild edible fungi more valuable than other NWFP, for example, and how do they compare in financial benefits with wood production? Some forests have a strategic as well as economic importance: they protect water catchments and fragile sloping land; they help to conserve biodiversity.

The challenge for planners and policy-makers is to balance the competing demands on forests and provide a framework within which forest managers can operate effectively. For wild edible fungi this means minimizing the impact of harvesting while allowing collectors fair and equitable access to forests; it means addressing the concerns of biologists who believe that commercial extraction is unsustainable while allowing local enterprises to develop. The sustainable production of wild edible fungi therefore has social, economic and even political dimensions.

Forest is used here in the general sense of areas where trees either occur naturally or are planted. The bulk of wild edible fungi harvests in terms of volume and value comes from species that form mycorrhizal associations with trees. Without the mycorrhizas the trees would grow poorly and the ecological integrity of forests around the world would be threatened. The impact of wild edible fungi harvesting should not disturb the mutual dependency of fungus and tree. The biology and ecology of wild edible fungi are therefore important, as is a fundamental knowledge of which species grow with particular trees. There are still many gaps in knowledge concerning edible ectomycorrhizal fungi and tropical tree species.

Forestry users include those who obtain wood products and NWFP (of which wild edible fungi are only one example). Forests also provide a range of services, some specific to particular users and others more generally valued. Ecological functions include protection of water catchments, erosion control and conservation of biodiversity. Forests provide social benefits, a place for leisure, sports and enjoying nature. The relationship between harvesting wild edible fungi and other products and services derived from forests needs to be understood and adjustments made to practices and management guidelines.

Decisions such as these depend on good data. There is widespread concern about unsustainable forest practices, including harvesting of wild edible fungi. This needs to be carefully examined using available data on yields, amounts harvested and other information about production. These topics are discussed later in this chapter.

Management of wild edible fungi has tended to concentrate on their biology and ecology, particularly those of high economic value. There is a considerable literature on truffles, for example (Federation-Française-des-Trufficulteurs, 2001), but few studies of edible species of *Russula* or *Lactarius*, many of which are collected and consumed locally in developing countries. Researchers are paying more attention to the complex

BOX 4

***Matsutake* and exports to Japan**

In Japan, *Tricholoma matsutake* is highly regarded and eating ceremonies are culturally important (Hall *et al.*, 1998a). Originally collected from Japan's forests, production declined steeply in the 1980s. The search for new sources identified American *matsutake* as an acceptable substitute (*Tricholoma magnivelare*) and it was quickly realized that substantial amounts could be harvested from the Pacific northwest of North America, where local use was minimal. The burgeoning trade with Japan coincided with a downturn in jobs in logging and timber extraction. Export businesses based on *T. matsutake* have also been established in Sichuan, China (Winkler, 2002; Yeh, 2000), Bhutan (Namgyel, 2000) and notably the Democratic People's Republic of Korea.

Exports of *T. magnivelare* and other closely related species occur from North Africa, Turkey and Mexico but details are sketchy. The amounts earned by these countries are small compared with Asia and North America. The prices paid by the Japanese vary considerably depending on the available supply each year and the quality of mushrooms when they arrive at market.

Matsutake is particularly valuable at an early stage of development and this requires careful searching in the upper humus layers of forests. Some collectors are not so careful: they rake the ground to uncover emerging fruit bodies, damaging the humus layer and affecting future harvests.

Matsutake is a mycorrhizal fungus and efforts have been made to "manage" natural ecosystems in the Republic of Korea and North America in an attempt to maximize production. Annual yields are still heavily influenced by available rainfall and ambient temperature at key times during the year.

(See Pilz and Molina (2002) for a general review of activities in North America.)

relationships between biological, social and economic issues, a welcome move towards establishing a sound basis for sustainable production of wild edible fungi.

Much has been written, relatively speaking, about *matsutake* (Box 4). This is an important export from several developing countries and there have been several accounts that examine the commercial harvesting in the wider context of forests and forest users (Winkler, 2002; Yeh, 2000). The Pacific northwest of north America is another area where management issues have been examined in detail (Pilz and Molina, 2002; Tedder, Mitchell and Farran, 2002). These studies are particularly useful in describing collectors and collecting practices and they provide a useful contrast to the few studies carried out for subsistence collections in developing countries (Lowore and Boa, 2001).

Concerns have been expressed about declining productivity and disappearance of certain species of macrofungi (Arnolds, 1995). Attention has focused on Europe and one of the identified issues was the impact of increased commercial picking in eastern Europe (Perini, 1998). Conservation of fungi is now an established topic of debate among mycologists. The debate has only just begun and it is important that it addresses the wider social and economic issues concerning harvesting if progress is to be made in halting the decline of any threatened edible species.

The following sections examine access to collecting sites, collectors and the impacts of harvesting. The chapter proceeds to an examination of published data on yields and production before attempting to provide practical advice on managing wild edible fungi for sustainable production.

REGULATING COLLECTION

There are widely differing rules and policies on the collection of wild edible fungi (see also Box 8, Chapter 4). Scandinavia has open access: anyone can pick edible fungi as long as they do not harm property (Saastamoinen, 1999). This policy has been challenged by economic migration from neighbouring countries, no longer part of the former Soviet Union, and the availability of cheap labour for collecting wild edible fungi and wild berries. Similar changes in eastern Europe have created new opportunities for commercial harvesting and led to concern about unsustainable harvests and how to regulate collections.

Controlling collectors is not always easy. After the Second World War the Finnish Government encouraged greater harvesting of wild edible fungi and continues to promote the use of an underutilized resource (Härkönen and Järvinen, 1993; Salo, 1999). Open access to the countryside is a tenet of life in Sweden and Norway and controlling the collection of wild edible fungi (and other NWFP) would require a fundamental change in national policies.

"Overharvesting" is a commonly expressed concern, both for commercial and subsistence collections. The fear among forest managers and others is that future production of wild edible fungi will decrease. These are genuine concerns but there is a danger of taking draconian steps to regulate collectors without understanding the impact of harvesting, based on an incomplete knowledge of how much is collected and what collectors do.

The main impetus for regulating collectors is where commercial harvesting occurs. The introduction of regulatory schemes serves a number of different functions:

- it attempts (in theory) to limit the amount harvested;
- it ensures that collectors are aware of best practice (least harmful picking methods);
- it provides income.

In Italy each province regulates who has the right to collect truffles (*Tuber* spp.). Collectors have to pass a simple test that confirms they are aware of how and where to harvest. Around 30 000 licences (each costing around US$90) were issued in Emilia Romagna in 2001 (Zambonelli, 2002, personal communication: *Truffles and collecting porcini in Italy*).

In Winema National Park, Oregon, the sale of permits provides a substantial income, though this is highly variable (Table 8). In Bhutan, only token amounts are earned from the sale of permits (Namgyel, 2000).

Local communities also administer permit schemes to limit access to valuable sites. This system appears to be less successful at reducing conflicts between neighbouring communities and problems have occurred in regulating collection of truffles in Spain (de Román, 2002, personal communication: *Trade in níscalos from North Spain to Catalonia and truffle production*). This is a reminder of the need to look closely at the fairness of schemes that unfairly exclude people rather than encourage equitable use of natural resources.

Collectors in developing countries frequently collect for subsistence uses and the edible fungi represent an important food resource. In Malawi, forest officers are concerned that allowing people to collect wild edible fungi in protected forest areas will lead to greater extraction of wood products, particularly firewood (Lowore and Boa, 2001). There is no officially registered commercial collecting in Malawi and there have been no attempts to introduce a permit system.

The success of regulation schemes depends on who controls or owns forests. It is a relatively straightforward matter to regulate collections of *Boletus edulis* in commercial pine plantations of South Africa compared to the more complex problems posed by multiple use of native forests in Malawi. The pressure to regulate access to sites comes from various sources, and not all involved in forestry. A strong conservation lobby in the United States has sought to limit commercial harvests (McLain, Christensen and Shannon, 1998).

The expansion of commercial harvesting in Europe has resulted in the introduction of regulations in Poland (Lawrynowicz, 1997); former Yugoslavia (now Serbia and Montenegro) (Ivancevic, 1997; Zaklina, 1998) and Romania (Pop, 1997). Information about the success of these schemes is sketchy and highlights the general difficulty of

TABLE 8
Sale of permits for collecting *matsutake* in Winema National Forest, Oregon, 1997–2002

Year	Permits sold	Value US$	End of season	Notes
1997	3 733	365 939	31 October	Biggest crop since 1989
1998	1 246	138 338	7 November	
1999	901	122 350	24 October	
2000	(512)	(61 180)	(21 September)	Data incomplete. No information after this date.
2001	not known	78 810	4 November	
2002	>1 200	>120 000	(4 October)	Interim data

Source: www.fs.fed.us/r6/winema/specialprojects. Commercial permits are valid for picking in the Deschutes, Umpqua, Willamette in addition to Winema National Forest. Only Winema publishes comprehensive accounts of the *matsutake* season (the "mushroom chronicles").

monitoring the conditions set by a permit. They often state how much can be collected in a fixed time but it is difficult to check this and collect penalties for transgressions.

Logging bans introduced in China (Winkler, 2002), the Philippines (Novellino, 1999), Canada (Tedder, Mitchell and Farran, 2002) and elsewhere have opened up new opportunities for collecting wild edible fungi and prompted concern about overharvesting. In Siberia, the opposite effect has happened: an increase in logging activities by foreign companies has made it more difficult for local people to collect wild edible fungi (de Beer and Zakharenkov, 1999).

Successful control depends on modifying regulations that do not work and maintaining a good dialogue with collectors (Pilz and Molina, 2002; see also Vance and Thomas, 1995). A pragmatic approach is needed to protect natural resources while allowing fair and equitable access to collectors.

COLLECTORS AND LOCAL PRACTICES

A recent study in Malawi describes what happened when Mr Kenasi Affad went collecting *bowa* (wild edible fungi) near his home in Machinga. He was accompanied by two researchers working for the Miombo Edible Fungi Project (Lowore and Boa, 2001).

"We set off at 6.00am, later than the normal time for start-off at 5.00am. Kenasi is equipped with nothing but the clothes he is wearing and a bucket. He is barefoot with no protection from the rain, which today is persistent but not heavy. He cannot afford to let the rain put him off as *bowa* collection is a rainy season activity and he must be prepared to get wet. This year the rains are still frequent and heavy which is good for the *kunglokwetiti* [6] and *chipatwe*.

He sets off on a well trodden path towards the places he knows where he shall find *bowa*. He has observed the rain for the past day or two, he knows what species are ready at this time, he knows where he went last time and the condition of the crop when he was last there. He uses all this information to decide where to go. These days – the end of the season – few *bowa* are found near to the home unlike early in the season when they are found in abundance.

At this time of year the main species found and the one preferred by customers is *kunglokwetiti*. These are found in rocky places and Kenasi has to be sharp to spot them. They appear here and there underneath droopy tufts of grass. To pick them Kenasi scoops the *bowa* from its base using his finger and gently lifts it from the earth. He then breaks the bottom part of the stem off and throws it away. He blows some of the remaining earth away and gently places the *bowa* in the bucket. He continues.

Kenasi knows that certain *bowa* are found near certain tree species and that each year the same type of *bowa* appear in the same places. He also knows that some species need a few days of rain followed by sunshine before appearing whilst others need prolonged rain. Some take a few days to emerge from a small fruit body to a harvestable *bowa*, others take a few hours. This is important because then he knows when to go back to the same place to look again for new *bowa*.

[6] *Cantharellus* species.

Kenasi shows us the path to Naiswe where he will go tomorrow. It will take about 3-4 hours solid walking to reach the place – then he can spend one hour collecting the *bowa* and come back within another two hours. It is normal for a collection trip to last up to six hours. Kenasi aims to fill a whole bucket (about 15 plates) before setting off for home. He always goes alone but may meet other collectors whilst in the forest. Passing on information about the whereabouts of *bowa* is sometimes done but there is not much point because it is simply a matter of chance – one might have missed what others will find. Kenasi will go collecting *bowa* from between 2 to 5 times a week, depending on the availability of *bowa* and customers.

In the past the eucalyptus were not there but there was indigenous woodland. *Bowa* were found in abundance just close to the village. Another reason why we have to travel so far these days is the number of people collecting. People simply want money so more and more people think of selling *bowa*. I can always find *bowa*, if the weather has been right, but it can take a long time to reach the place and a long time to fill a whole bucket."

This short account graphically describes the type of problems that a collector has to cope with. Kenasi knows where to look though he also knows that he has to be lucky to make a good collection. He comments on the loss of native woodland, where the fungi are most abundant, and he says that he must travel further to collect wild edible fungi because now there are more collectors.

Kenasi lives close to the forest and is part of a community that depends on the miombo woodland for food, income and shelter. Collecting *bowa* is an important source of income for him but it is only one way of earning a living from the miombo. Increasing numbers of people have taken the opportunity to collect, as Kenasi observes, because in the area where he lives there is a good selling point on a major road near to the forest.

Kenasi is unusual because the collectors in Malawi are mostly women, as is the case in the United Repubiblic of Tanzania (Härkönen, 2002) and Burundi (Buyck, 1994b). Table 9 describes collectors and their practices in a number of different countries. In China most collectors are men. Both men and women are involved in Mexico, where there is extensive harvesting each year. In Malawi the maximum time taken for collecting wild edible fungi and getting them to market is less than 24 hours. Any longer and the mushrooms for sale deteriorate and are worth much less. Women in Mzimba district in northern Malawi walk up to 10–15 km to get to the nearest market in Mzuzu. This limits collecting to a six hour collecting trip (there and back) from their homes (Lowore, Munthali and Boa, 2002). Distances from house to forest to selling points are shorter in Liwonde, near Zomba (Lowore and Boa, 2001) because of the proximity of a main road, a common selling point for wild edible fungi in several African countries (Plate 6).

In the Russian Federation and Ukraine whole families go on collecting trips and these appear to be more of a social event than collecting in order to sell. The distances travelled to the best sites can be substantial (Table 9). Immigrants collect wild edible fungi in the Klamath bioregion (northern California), many of southeast Asian origin (Richards and Creasy, 1996), attracted by job opportunities. They soon realize that competition is fierce and that incomes are not guaranteed. There have been some clashes between collectors and a general suspicion of people from southeast Asia, partly because of their poor English and a failure to observe regulations about where to pick. An account by an American picker of *matsutake* (Moore, 1996) provides a personal account of some of the antagonism that migrant labour may have to overcome – successfully overcome in this particular case.

Where money is involved in collecting wild edible fungi problems may arise, sometimes fuelled by exaggerated stories of potential earnings. Villages in Sichuan engaged in sustained battles to determine local rights to *matsutake* sites culminating in the sabotage of water supplies – they were without water for 45 days – and destruction of a key bridge. One village threatened not only to continue their disruption of life in

TABLE 9

Collecting wild fungi in the United Republic of Tanzania, Mexico, the Russian Federation, Bhutan, Finland, India and China

Activity/issue	United Republic of Tanzania
Who collects?	Mainly women and children though men bring them home if they happen upon them.
Collecting	Travel by foot to sites. Open access. No special harvesting methods are used and official regulation of collectors is absent. People go out early to collect because of competition for edible fungi – hinting at the importance of selling in local markets.
Local traditions, choice of species	Elderly country people whose families had lived in the same place for several generations knew most about wild fungi. Many more species eaten in miombo areas than hills. Boletes eschewed by all: "even monkeys won't eat them" (monkeys eat *B. edulis* in Malawi, however). People were well aware of poisonous varieties. Some groups of people will not eat any wild edible fungi. Educated people have forgotten almost everything about wild fungi. A similar diminishing of local tradition can be found in Malawi and Zimbabwe.
Activity/issue	**Mexico**
Who collects?	Families and individuals of both sexes. Photos of market places show only women selling.
Collecting	Collectors walk 4–5 km a day, carrying around 4–5 kg to be sold in 5–7 hours. Collections transported up to 55 km; not clear if this is done by traders and/or collectors. Open access to sites. There are government regulations for picking seven major species.
Local traditions, choice of species	All types of macrofungi are collected. Long tradition of wild fungi use. Knowledge lost as people move from rural to urban areas; acceptance of wild fungi may dwindle especially as availability of cultivated species increases. Generally low frequency of poisoning cases.
Activity/issue	**Russian Federation [Siberia]**
Who collects?	Families.
Collecting	5–6 km from boundaries of village or public transport stops. Some drive 40–60 km. No restrictions on access to sites, except nature reserves and national parks. Daily harvest could be from 15 to 100 kg per person in good years.
Local traditions, choice of species	Long history of collecting which has intensified with worsening economic situation. More people unable to afford imported food while food distribution within the Russian Federation has declined. Also, reduced employment opportunities in mining and forestry industries. 18–25 species are regularly collected; *Lactarius deliciosus* and Boletus edulis most important. Poisoning incidents not noted separately for this region but see Table 5 for reports from other parts of the Russian Federation.
Activity/issue	**Bhutan**
Who collects?	Families.
Collecting	On foot. Some camp out and begin collecting with torches very early in the morning because of competition. Local farmers do not allow farmers from other *geogs* to visit their area. The National Mushroom Centre has provided training on sustainable harvesting to 1 525 farmers. Concern expressed about damage to *matsutake* mycelium in soil because of harvesting methods.
Local traditions, choice of species	Little known about tradition of wild edible fungi but thought to be well established. Attention now focused on *matsutake* which had a low, local value until exports to Japan began.
Activity/issue	**Finland**
Who collects?	No gender or age differences noted.
Collecting	Collectors travel by public and private transport to sites. Open access except peoples' back yards. Collection is actively encouraged following inventory which shows that only a small proportion of the wild edible fungus resource is used each year.
Local traditions, choice of species	Official advice provided on best fungi to collect, originally because of famine conditions and later seeking to encourage best use of wild food resources. Western Finland favours different species to Karelians in East, whose tradition of collecting and eating is much stronger.
Activity/issue	**India [Madhya Pradesh]**
Who collects?	Whole families involved but women more active.
Collecting	Tribal people well acquainted with habitat and period of fruiting. No restrictions on access to collecting sites are mentioned.
Local traditions, choice of species	Several species are collected.

ACTIVITY/ISSUE	CHINA [YUNNAN]
Who collects?	Men are more interested in collecting.
Collecting	People do not go collecting on a regular basis because cultivated species are available throughout the year.
Local traditions, choice of species	Only mountain areas are visited; highest number recounted by one man was 33 edible species. People well aware of poisonous species.
ACTIVITY/ISSUE	**CHINA [SICHUAN AND ALLIED AREAS]**
Who collects?	Not stated.
Collecting	Most concern about declines in *matsutake* production is for Degen Tibetan Autonomous Prefecture in northwest Yunnan. Has the highest extraction rates with clear decline in productivity. This is linked to bad harvesting techniques (raking). When sold by size encourages damaging harvest methods. No decline in productivity in Litong's Jumba valley where sold by weight. Collectors of *Cordyceps sinensis* in Litang County are confined to legal grazing grounds or to forests where they have right of access. Outsiders must pay a fee to local community for collecting and clashes have occurred. Collection of other edible species is widespread (Rijsoort and Pikun, 2000).
Local traditions, choice of species	Long tradition of collecting edible and medicinal species. *Matsutake* not commonly collected before 1988.

Sources: UNITED REPUBLIC OF TANZANIA – Härkönen, 2002; MEXICO – Bandala, Montoya and Chapela, 1997; Montoya-Esquivel *et al.*, 2001 and www.semarnat.gob.mx. RUSSIAN FEDERATION – Vladyshevskiy, Laletin and Vladyshevskiy, 2000; BHUTAN – Namgyel, 2000. FINLAND – Härkönen, 1998; Pekkarinen and Maliranta, 1978; INDIA (MADHYA PRADESH) – Harsh, Rai and Soni, 1999. CHINA (YUNNAN) – Härkönen, 2002; CHINA (SICHUAN and allied areas) – Winkler, 2002; Yeh, 2000.

the rival village but to "hide the pieces of the water pipes in the forest so that they could not be repaired" (Yeh, 2000). Such conflicts are unusual but when money becomes the main motive for collecting, management of collectors (and access to sites) needs careful adjudication.

Most collectors work alongside each other without any obvious problems. This does not mean that they necessarily cooperate in harvesting. In northern Spain, *Lactarius deliciosus* (níscalos) are sold to buyers from Catalonia, earning small but useful amounts of money. Even close friends refuse to reveal the location of favourite sites (de Román, 2002, personal communication: *Trade in níscalos from North Spain to Catalonia and truffle production*).

Commercial collection of wild fungi is a recent and small-scale activity in Scotland. Previously there was sporadic and minor picking for personal use. Landowners of the mostly private forest areas involved expressed a number of concerns about the influx of collectors (Dyke and Newton, 1999):

- unauthorized access by collectors to their land;
- lost revenue: the owners did not benefit from the collections on their land; they were also unable to earn money from organized fungal forays if the mushrooms had already been picked;
- damage to resource (wild edible fungi and the forest);
- conflicts with hunting (an important source of revenue for some landowners).

A total of 53 percent of collectors interviewed in Scotland did not know who owned the land they collected from. This study is a good example of how to collect information for developing management plans.

Collectors come from a wide range of social classes but the overall impression is that the majority are poor rural people who have traditionally lived close to the land and for whom wild edible fungi are a common and often unrecorded source of food (De Kesel, Codjia and Yorou, 2002).

HARVESTING METHODS AND APPROACHES

Harvesting

The impact of harvesting wild edible fungi is frequently raised and a recent review provides a helpful summary of key issues that are explored in further detail below (Pilz and Molina, 2002).

Collecting wild edible fungi is often compared with picking fruit from a tree. Removing all the fruit does not affect future harvests unless the tree is damaged, but might have an impact on regeneration. This appears to be true for wild edible fungi but with some reservations: removing unopened fruiting bodies prevents dispersal of spores. In some areas of Italy regulations prevent the collection of first flush of some edible species (Zambonelli, 2002, personal communication: *Truffles, and collecting porcini in Italy*). (This makes practical sense too, since the early fruiting bodies are often damaged by insects.) Some collectors spread parts of the mushroom cap to encourage dispersal of spores.

A study in Switzerland showed that harvesting all the fruiting bodies of 15 species of macrofungi over a ten-year period had no significant effect on production (Egli, Ayer and Chatelain, 1990). If soils are compacted or leaf litter layers are disturbed, this can affect production. Indiscriminate digging for truffles, for example, is harmful. Crude raking to reveal young and immature *matsutake* damages the mycelium present in the upper layers of the soil. (The young fruiting bodies can be sold for a higher price.) This can be avoided by first identifying potential areas of *matsutake*, then using your hand to locate the tell-tale bumps while generally looking for signs of emerging fruit bodies (Arora, 1999).

Most species of edible fungi are picked without causing any damage since their fruiting bodies and edible parts are all above ground. The search for truffles (*Tuber* spp.) is often undertaken by trained dogs (Plate 4) (Hall *et al.*, 1998a). The traditional use of pigs is now banned in Italy because they are difficult to control and sometimes eat the truffles. Truffle dogs are not used in China and random digging used to locate fruiting bodies will affect future production.

The Swiss study also showed the effect of trampling on the production of one chanterelle species. However, "normal" yields were restored once the trampling stopped (Egli, Ayer and Chatelain, 1990). Trampling is not thought to be a common source of damage. The number of collectors per unit area of forest is usually low and there is no evidence that trampling has affected yields in Malawi, for example. Commercial harvesting does increase the pressure on sites though wild edible fungi usually occur over a wide area and collectors keep apart in their searches.

Enhancing productivity

The decline in *matsutake* production in Japan in the 1980s prompted research on how to maximize yields *in situ*. Some success was achieved, although the increases in production failed to stem the overall decline. In the Republic of Korea methods included watering and vegetation control (Koo and Bilek, 1998). In Finland, soil surface treatments were examined for enhancement of the production of *Gyromitra esculenta* (Jalkanen and Jalkanen, 1978). These approaches are potentially costly and it is not known how successful they have been in increasing financial returns.

An alternative is to manage forests in a way that increases production of wild edible fungi. Attempts have been made in the Pacific northwest of North America to balance the production of wood and wild edible fungi (Weigand, 1998). The conclusions of a study of management of native stands of conifers in the United States and the production of wild edible fungi, including *Tricholoma matsutake* and chanterelles, are summarized below (Pilz and Molina, 2002):

- Clear-cut harvesting disrupts the production of most edible ectomycorrhizal fungi for ten or more years. It only recovers once the fungi have re-established on trees that are old enough to provide necessary nutrients.
- A thinned stand (one where trees are selectively removed to encourage growth of remaining trees and to remove weak specimens) introduces more rain and sunshine and more rapid wetting and drying of the forest floor. Heavy thinning at one site of Douglas fir reduced chanterelle fruiting by 90 percent in the following

year. Less frequent thinning might help to maintain fungal productivity but the loss of wood production might outweigh the benefits.

- Compaction of soil from logging operations reduces productivity while the removal of large branches makes it easier and safer to find wild fungi without necessarily increasing base productivity.

The critical issue in enhancing production of wild edible fungi is their economic importance compared to the value of wood production and other forest uses. This is often poorly understood because accurate data are missing on the value of harvests.

MEASURING PRODUCTION

Yields

Data from experimental studies in five countries are summarized in Table 10. Comparisons are difficult to make because some studies include all edible species while others measure the productivity of individual species. Sampling methods also vary, with plot size and total area monitored often too small to draw any major conclusions.

The results from Mexico suggest that up to 1 759 kg per ha of wild edible fungi can be produced in a good year. Yields from other countries are usually much lower, around 100 kg per ha and less. Natural fluctuations occur from year to year (Villarreal and Guzmán, 1985; Villarreal and Guzmán, 1986a; 1986b) and without historical data it is difficult to draw any useful conclusions from a single year's production. There is a clear need to improve the quality and range of data on yields. Concerns have been expressed about "declining yields" yet there is also a lack of published data that allow a closer examination of the impact of commercial collecting in Portugal (Baptista-Ferreira, 1997) and the Russian Federation (Kovalenko, 1997), for example.

TABLE 10
Yields of wild fungi from different countries

Country	Details of annual yields	Amount (kg/ha)	Source
Russian Federation (central Siberia)	"Most popular (edible) mushrooms"	65–170	Vladyshevskiy, Laletin and Vladyshevskiy, 2000
Russian Federation (Arkhangelsk)	(a) *Lactarius torminosus*, (b) "red-headed mushroom" - ?*Russula* sp.	(a) 2–14 (b) 9	Chibisov and Demidova, 1998
Finland (north)	All edible mushrooms at Sotkamo (a) 1976 and (b) 1977	(a) 30 (b) 85	Koistinen, 1978
Finland	*Gyromitra esculenta* (note fluctuations; 1973 and 1974 good; 1975 and 1976 poor; 1977 mediocre)	50–100	Jalkanen and Jalkanen, 1978
Estonia (northwest)	Average for all edible fungi at three sites, from 1978 to 81 *	124, 499,143	Kalamees and Silver, 1988
Estonia (northwest)	Average for (a) *Suillus variegatus* – one site and (b) *Lactarius rufus* – three sites *	(a) 41 (b) 20; 24; 405	Kalamees and Silver, 1988
Mexico	All edible species from two sites	85	Lopez, Cruz and Zamora-Martinez, 1992
Mexico (Veracruz)	All edible species, two sites (a) and (b) for 1983 and 1985 respectively	(a) 1 759; 234 (b) 747; 180	Villarreal and Guzmán, 1985; 1986a
Mexico (Veracruz)	(a) *Suillus granulatus*; (b) *Cantharellus cibarius* (c) *Amanita caesarea*; (d) *Boletus edulis* For 1983 and 1985 respectively	(a) 246; 75 (b) 4; 8 (c) nd; 38 (d) 150; 9	Villarreal and Guzmán, 1985; 1986a
United States (Pacific northwest)	(a) *Tricholoma magnivelare*; (b) *Morchella* spp.; (c) *Cantharellus* spp.	(a) 3–15 (b) 1–6 (c) 2– 0	Pilz and Molina, 2002

Amounts are fresh weight or presumed to be so. Villarreal and Guzmán data based on extrapolation from two permanent plots of 100 m^2 at each site.
* Insect damage reduces available harvest of non-*L. rufus* edible species by around 70 percent. nd – no data.

TABLE 11
National production of wild edible fungi

Country	Item (wild edible fungi)	Amount (tonnes)	Source
Belarus	"Resources" from 1981 to 1985	53 000	Malyi, 1987
Canada	Estimated annual export	220–450	de Geus, 1995
Canada	Chanterelles, boletes and morels "exported in a good year"	1 000	Wills and Lipsey, 1999
China	Production of boletes, *Lactarius deliciosus* and "others" (?wild edible fungi): 1998	308 000	Sun and Xu, 1999
Estonia	Average annual export 1929–38	2 200	Paal, 1999
Finland	Yields in (a) 1988, (b) 1992 and (c) 1996	(a) 1 050 (b) 670 (c) 360	Härkönen, 1998
Poland	Production of (wild) edible fungi in 1958	3 500	Bukowski, 1960
Russian Federation (Arkhangelsk)	Collected annually by local people in 1930s	2 040	Chibisov and Demidova, 1998
United States (WA, OR, ID)	All wild edible fungi collected for trade: 1992	1 776	Schlosser and Blatner, 1995

Amounts are fresh weight or presumed to be so in the absence of other information. Production data from boreal and cold temperate countries, e.g. Lithuania, were seen too late to be included in this table (Lund, Pajari and Korhonen, 1998). See Chapter 4, section: *National and international trade*, for related information on the value of wild useful fungi (edible and medicinal).

Table 11 summarizes national data on the amounts harvested of mostly commercial species. Total production in any given period will be higher. Data for developing countries are poorly represented and an attempt has been made to estimate the potential production for Tlaxcala state in Mexico, where wild edible fungi are widely collected. Tlaxcala has 83 000 ha of forest of which 65 000 ha are conifers and broadleaves. The remaining area has only broadleaf species. A potential yield of 10 kg per ha per year for all wild edible fungi in the 65 000 ha would provide a potential harvest of 650 tonnes. One of the main, if not principal, limiting factors in how much is harvested and sold is the time taken to collect and bring the fungi to a potential buyer.

The important question of how much of the total production is actually harvested in any one year remains largely unanswered, even for commercial species of wild edible fungi.

Inventory

Concerted efforts have been made to estimate productivity of commercial species of wild edible fungi in North America (Pilz and Molina, 2002). Similar approaches were used in Malawi to monitor production of edible species (Meke in Boa *et al.*, 2000). A total of 250 50 m × 2 m plots were assessed at five native (miombo) woodland sites from 1999 to 2002 and initial results are available at www.malawifungi.org. Information collected included the number and weight of fruiting bodies and their proximity to trees (to examine mycorrhizal associations).

Fruiting bodies of macrofungi appear over a potentially large area and one recommendation for collecting yield data is to use long, narrow plots, as noted above. This also minimizes trampling damage by field staff. The frequency of observations depends on when particular species appear. Local collectors have proved a helpful source of information in Malawi.

More and better data are needed on yields and productivity to assist in drawing up management plans. Further advice on methods for assessing production of NWFP have been published by FAO (2001a).

Market surveys provide a guide to general productivity and are a simpler and less costly way of collecting data, provided that significant amounts are sold to the public.

PRACTICAL PLANNING: TOWARDS SUSTAINABLE PRODUCTION

The ultimate aim of managing wild edible fungi is to achieve sustainable production. The importance of good quality data has been emphasized and attention drawn to general issues of forest management and forest users. The first steps in formulating a management plan are to describe and then analyse the features of each production system. Table 12 suggests a general approach to adopt with key questions to ask.

Finland is a rare example of a country that has actively attempted to manage its wild edible fungi resources. They have actively supported wild edible fungi (together with wild berries) since the Second World War and their widely published experiences provide helpful pointers for other countries. Mexico has also shown a sustained interest in managing wild edible fungi. Coordinated efforts have been made by researchers and local and regional government to understand the importance of wild edible fungi and manage them for the benefit of people and the environment.

Much of the information required to begin the management planning process is already available in countries such as China (Mao, 1998) and Turkey (e.g. Gurer, 2002, personal communication: *Unpublished data on wild edible fungi for Turkey*). The former Soviet Union devoted much effort to investigating wild edible fungi (Paal, 1998), although perhaps more from the viewpoint of the fungi than their social and economic related features. This is a general weakness in many countries and an area where particular efforts are needed to improve knowledge.

Fair and equitable access to forests and forest resources is a critical issue. If people believe they are unfairly excluded they may continue to collect but not observe

TABLE 12
Preparing management guidelines for wild edible fungi

TOPIC	QUESTIONS TO ASK/KEY POINTS
Ownership of forests	***Public or private?*** State/region controlled or under joint management with rural people? As the number of stakeholders increases so the task of resolving who has user rights and how these are moderated becomes more complex. Private owners may be unaware of the value of WEF and this should be carefully explained so that they have realistic expectations about financial returns from potential commercial operations.
Relative importance of wild edible fungi	***Commercial or personal?*** Firstl consider the value of WEF by themselves and then compare this with other forest products and services. Review all WEF species together for preliminary assessments and later look more carefully at the value of different types (which may vary significantly). Personal collections include subsistence and recreational uses (e.g. amateur mycologists, field biologists). Good and reliable data on production and amounts harvested are essential for effective planning. If these data do not exist or are patchy, consult collectors to assess patterns of previous use and consider an inventory based on a system of sample plots.
Collectors and their practices	***People profiles and harvesting methods.*** Who are the collectors: are they local or hired labour brought in from other places? Examine the harvesting practices and assess their impact on WEF resources and the forest and trees. Review the need to change practices and how collectors could be encouraged to use less harmful methods. Look carefully at the other features of collector livelihoods so that WEF can be put in wider contexts.
Legislation and regulations	***Collecting permits and right of access.*** How are collections of WEF regulated and do the current laws support sustainable use? The key principle is fair and equitable access to forest resources which maintains a healthy balance between use of WEF and other forest uses. Examine current legislation to see whether it is enforceable and reflects current needs of users. The guiding principal is pragmatism: regulations that work.
Production and financial value	***Volume and value.*** Assess this on a national scale since data will be used to develop government policies. Weak data lead to weak policies and management of WEF has been hampered by wrong perceptions and knowledge of collecting practices and their importance, particularly to rural communities.

BOX 5

Practical inventory: experiences from Malawi

An extensive review of inventory studies for NWFP has revealed the poor quality data that often emerge at the end of studies, and highlighted the general paucity of information on productivity (FAO 2001a). This is a critical issue if foresters are to understand the impact of harvesting practices on wild edible fungi and to resolve the competing claims of commercial interests and other groups that have an instinctive suspicion of collectors (which often includes the foresters themselves).

In Malawi, enumerators were hired in to collect data from four major sites. There were few major problems apart from the failure of data collection at one site which was resolved the next year when a local non-governmental organization (NGO) helped. It took at least one season for all concerned to become familiar with protocols and techniques. The rains were poor in the second and third year and productivity consequently low. A good knowledge of local and scientific names of wild edible fungi was a major benefit in interpreting the data.

The cost of travelling to the four sites was high; fuel is expensive in Malawi and budgets should be calculated before finalizing the location of plots. There may be little advantage in travelling afar unless these sites are significantly different from those closer at hand. A computer data entry system was created at the very beginning and was invaluable in allowing yield data to be entered swiftly and accurately. It soon became clear if wrong data had been entered or if there were unexplained gaps. The supervisors used this information to suggest improvements in how enumerators collected data and reported the results.

Analysis of the data and drawing conclusions have proved more difficult to achieve, partly because the people involved in the work worked far apart and data collection was continued up to the end of the project. It would have been better, in retrospect, to stop data collection earlier and to give a longer period of time (six months) for data analysis.

More could be done to provide practical advice on how to take inventories of wild edible fungi. There is a lot of useful information available on NWFP (FAO, 2001a), but there is as yet no simple, practical guide that would encourage more people to measure productivity and show them how to perform simple analyses of the data.

regulations or pay permits or taxes. People routinely avoid paying official taxes in Italy when harvesting *Boletus edulis* and truffles (Hall *et al.*, 1998b). Exclusion can also turn to resentment. In northwest Spain in 2001, a truffle site was crudely raked overnight and "spoiled" for collection because a previous resident of an area was no longer allowed to obtain a collector's permit for his former village (de Román, 2002, personal communication: *Trade in níscalos from North Spain to Catalonia and truffle production*).

The Scottish Wild Mushroom Code[7] provides the following guidelines to collectors of edible and non-edible species:

- only pick what you will use; wildlife need mushrooms too;
- do not pick until the cap has opened out, and leave those that are past their best;
- take care not to damage the main part of the mushroom below the surface and not to damage its surroundings;
- scatter trimmings discreetly in the same area the mushroom came from;
- only pick what you know and take a field guide to identify mushrooms where you find them; some mushrooms are poisonous and rare ones should not be picked;
- please observe special conditions that may apply to nature reserves.

Codes of practice are useful but again must be realistic if they are to be adopted.

The loss of native forests reduces the potential production of wild edible fungi. Planting exotic trees opens up new possibilities, some of which have already been

[7] Available at: www.rbge.org.uk/research/celtica/fungi/sustainability.htm.

exploited. *Boletus edulis* has been introduced to South Africa and a small export trade has been established (Pott, 2002, personal communication: *Export of* Boletus edulis *from South Africa*). This fungus is not eaten locally. A eucalypt species from Australia, planted in Madagascar, has formed mycorrhizal associations with a "native" edible *Russula* (Buyck, 2001). Similar interactions involving other wild edible fungi have been observed in West Africa (Ducousso, Ba and Thoen, 2002).

Planting exotic species does not, therefore, necessarily impoverish the local mycota (Ryvarden, Pierce and Masuka, 1994) and may significantly increase opportunities for collecting WEF, as has happened with the planting of *Pinus nigra* in northwest Spain and the commercial markets for *Lactarius deliciosus* that have developed over the last thirty years (de Román, 2002, personal communication: *Trade in níscalos from North Spain to Catalonia and truffle production*). New Zealand has seized the opportunity to introduce edible mycorrhizal fungi, and the lack of competing native species of fungi is seen as a positive opportunity in support of commercialization (Hall and Wang, 2002).

PLATE 4

TRUFFLE COLLECTING IN ITALY

The collection and cultivation of *Tuber* spp. is of commercial importance. Truffle photographs are from Urbino, Marche in Italy, and are of *Tuber aestivum* unless otherwise stated. All photos by Eric Boa.

4.1 Luna uncovers the truffles and awaits a reward. Dogs are easier to handle and cause less damage than pigs

4.2 Pierluigi displays the truffles after digging them up with the long-handled tool. It has a curved blade at the end.

4.3 The clearing is a truffle "orchard", *tartufaia* (It.) or *truffière* (Fr.). Trees are infected artificially with the fungus.

4.4 Marking the test taken by truffle collectors in Bologna to confirm they know how and where to harvest.

4.5 *Tuber aestivum*, cut open to show distinctive flesh.

4.6 The suppressed vegetation (*brulée*), suggests *Tuber aestivum* is present.

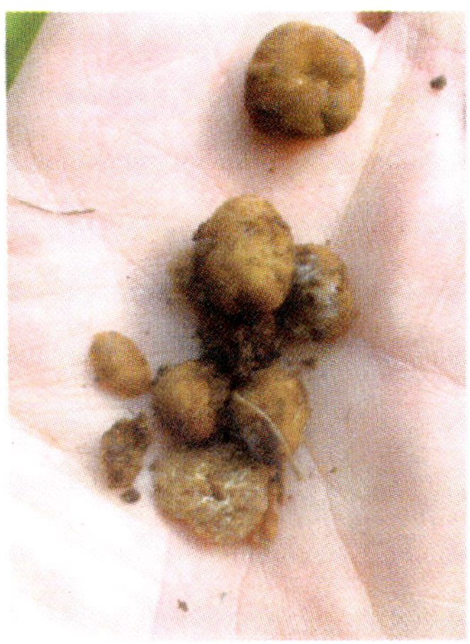

4.7 *Tuber excavatum*, largely worthless. Not all truffles are equally prized. Present at the same site as *T. aestivum*.

4.8 Some truffle collectors raise and train their own dogs. Elvisio also sells to other collectors.

4.9 *Tuber magnatum* for sale as a luxury food, costing around US$35 per jar.

PLATE 5
THE TRADE IN *BOLETUS EDULIS*

These valuable and sought after wild fungi grow around the world yet are not eaten in countries such as Malawi. The trade is dominated by Italians, both at home (factories) and overseas (as traders). Huge volumes are imported from China, eastern Europe and southern Africa. Known in Italian as *porcini*, they are dried and sold preserved, sometimes in mixtures with other *Boletus* species and other cultivated mushrooms. All photographs from Borgo Val de Toro, Parma, Italy, unless stated otherwise, and taken by Eric Boa.

5.1 *B. edulis*: produced in abundance yet not eaten or collected. Pine plantation, Zomba plateau, Malawi.

5.2 Fresh *porcini* being prepared for cooking and preservation in brine, prior to being sold.

5.3 *Porcini* cooked and ready for bottling..

5.4 Preparing jars of *porcini* and other mushrooms.

5.5 A range of mushroom products, including chanterelles and paddy straw.

5.6 Dried and preserved *porcini* on sale.

5.7 Permits are required to collect wild fungi in this valley. Residents and property owners pay less compared to "outsiders".

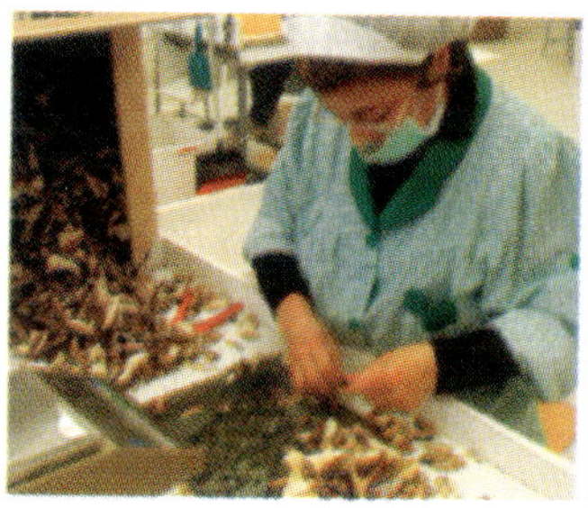

5.8 Dried *porcini* from several countries are carefully graded.

5.9 *Porcini* and other mushrooms in brine, as imported from overseas.

5.10 Other species of *Boletus* are sometimes mixed with *porcini* and sold.

5.11 *Pholiota nameko* from China, also sold in mixtures with *porcini*.

5.12 Young specimens of *porcini* in brine.

4 Importance to people: food, income, trade

WILD EDIBLE FUNGI AND LIVELIHOODS

This chapter looks at the ways in which wild edible fungi are important to people, particularly those in developing countries, and attempts to relate this information to the way in which people live. Development support is adopting new approaches towards helping poor people in devloping countries. Pragmatic and practical approaches to reducing poverty seek improvements sooner rather than later. Wild edible fungi already play an important role in the lives of many people and more benefits could be achieved. A knowledge of the fungi themselves is important but will not of itself lead to changes unless the choices and options defined by livelihoods are closely examined (Box 6)[8].

Wild edible fungi provide two main benefits to people: they are a source of food and income. Around six percent of edible species also have medicinal properties (next section; Table 14). This contribution to human welfare is difficult to assess and has received little attention. The medicinal properties of mycorrhizal fungi have not been well investigated (Reshetnikov, Wasser and Tan, 2001).

The awareness of wild edible fungi and their importance to people are generally poor. Subsistence uses in developing countries have often been ignored and it is only in recent years that initiatives on NWFP have begun to explain their widespread use and roles in livelihoods. There has been much interest in the last years surrounding commercial harvesting of *matsutake* in the Pacific northwest of North America, supported by a substantial literature. However, *matsutake* and the continued interest in truffles and truffle cultivation (Hall, Zambonelli and Primavera, 1998) reflect a very different pattern of use, where wild edible fungi are seen as a luxury food.

Beyond the glare of publicity of commercial harvests, information from development projects and national initiatives – for example China, Mexico and Turkey – has slowly been emerging. Commercial harvesting also benefits rural people in several countries but the sum of the money earned is less than the total benefits gained from widespread subsistence uses. Substantial benefits are derived by people in developing countries, and in particular the most vulnerable communities living in rural locations – the "poor of the poor".

Global statistics are not available and the evidence to support statements about widespread benefits is based first on case studies, discussed in more detail below, and second on more anecdotal accounts. Information has been poorly documented in the past because of fewer opportunities for scientists to study wild edible fungi in developing countries. There have also been cultural biases against wild edible fungi and an often unjustified assumption that they are of minor importance (Piearce, 1985; Wasson and Wasson, 1957). The latter publication has done much to stimulate wider interest and more research (Table 13).

Donor-funded projects on wild edible fungi in the United Republic of Tanzania (Härkönen *et al.*, 1993), Malawi (Boa *et al.*, 2000) and Benin (De Kesel, 2002, personal

[8] See www.livelihoods.org for further information and explanations of what is known as the sustainable livelihoods approach. Throughout this chapter livelihoods is used in the broad sense of the means by which people live.

BOX 6

Development projects and wild edible fungi

Two different approaches to wild edible fungi are compared. In the first hypothetical project, all the species of wild edible fungi in a region are described and nutritional characteristics are analysed. Local names are gathered and general observations made about local marketing.

In the second project, researchers assess sources of food and income for local communities. They compare their relative importance and examine the opportunities and constraints to improved nutrition and income, which include edible fungi. New schemes and initiatives are agreed and piloted.

The two approaches are complementary but the first project does not lead to change in local practices. The second project seeks to make improvements to the way people live based on available information. More improvements might be achieved if better technical knowledge was available, yet local communities can still plan new initiatives using local names for wild edible fungi or seek efficiencies in local marketing based on a clear understanding of local practices and opportunities.

communication: *Wild edible fungi from Benin*) have taken a broader view of social and economic issues related to wild edible fungi. National programmes in Mexico have established a sound knowledge of the many species of wild edible fungi used throughout the country (Villarreal and Perez-Moreno, 1989). Research attention is now being turned on social and economic factors, encouraged by a wider awareness of the importance of NWFP to rural economies and people.

The importance of wild edible fungi to people in developing countries may also have gone unremarked for the simple reason that many of the collections are for personal use (Yorou and De Kesel, 2002). The limited mycological expertise in West Africa is said to be responsible for the mistaken belief that it is a "mushroom desert" (Ducousso, Ba and Thoen, 2002). Reports from Ghana (Obodai and Apetorgbor, 2001) and Sierra Leone (Down, 2002, personal communication: *Wild edible fungi from Sierra Leone*) indicate that local use is widespread. The regular use of wild edible fungi in tropical rain forests was revealed when careful observations of local practices were undertaken in Brazil (Prance, 1984), now supported by evidence from Kalimantan (Leluyani, 2002, personal communication: *Edible fungi of Kalimantan*) and Sarawak (Chin, 1988; Jones, 2002, personal communication: *Wild edible fungi use in Sarawak*).

Information is published in a number of different places or disciplines (Table 2) and is sometimes presented in broader studies of communities (e.g. Shackleton *et al.*, 2002: South Africa; Ertrug, 2000: Turkey; Gunatilleke, Gunatilleke and Abeygunawardena, 1993: Sri Lanka). These and many other reports listed in the reference section emphasize that the contributions of wild edible fungi to diet and income of rural people should not be underestimated.

The following sections take a closer look at the types of benefits obtained from wild edible fungi. Their relative contributions to livelihoods vary greatly. A meal of wild mushrooms is a delicacy in Switzerland or the United States but a necessity in Malawi. The money earned from selling *Lactarius deliciosus* provides a small financial fillip in northern Spain (de Román, 2002, personal communication: *Trade in níscalos from North Spain to Catalonia and truffle production*) while collecting morels in India allows people to pay for sending their children to school (Singh and Rawat, 2000).

The importance of wild edible fungi from a development perspective is defined by comparison with other sources of food and income. Alternatives do exist and proposals to increase the use and benefits of wild edible fungi will always be compared with available options. The lure of jobs in the tourist trade in Hunan, China, is an attractive alternative to climbing up and down mountains, with no guarantee of finding wild edible fungi to sell (Härkönen, 2002). The contraction of job opportunities in the

TABLE 13
Ethnoscientific studies of wild fungi with edible and medicinal properties

COUNTRY	WILD FUNGI EMPHASIS	SOURCE
Australia	Useful (includes edible) species in aboriginal culture	Kalotas, 1997
Balkan region	Medicinal species: study of eastern Slavs	Didukh, 2001
Brazil	Study of Sanama Indians (includes edible species)	Fidalgo and Prance, 1976
Canada	Aboriginal plant use, including edible and medicinal wild fungi	Marles *et al.*, 2000
China	Comparison of Hunan and China (mostly edible species)	Härkönen, 2002
Guatemala	Folklore concerning *Amanita muscaria*	Lowy, 1974
General	Fungi in folk medicine	Birks, 1991
General	The origins of ethnomycology, as a discipline	Davis, 2000
General (Mexico)	Personal stories of ethnomycology, myths and ceremonies	Riedlinger, 1990
Himalaya, eastern	Edible fungi of medicinal value	Boruah and Singh, 2001
India	Fungi in folk medicine	Vaidya and Rabba, 1993
India, central	Ethno-myco-medicinals	Rai, Ayachi and Rai, 1993
Japan	Uses of fungi and lichens by Ainu	Yokoyama, 1975
Malawi	Edible, medicinal and species used for ceremonial purposes	Morris, 1992
Mexico	Medicinal mushrooms: traditions, myths and knowledge	Guzmán, 2001
Nepal	General observations (mainly edible species)	Adhikari and Durrieu, 1996
Nigeria	Medicinal practices in Yoruba culture	Oso, 1977
Papua New Guinea	Mainly concerned with edible species	Sillitoe, 1995
Peru	Fungi, mostly edible, part of ethnobotanical study	Franquemont *et al.*, 1990
Poland	Polish folk medicine	Grzywnowicz, 2001
Russian Federation	Khanty folk medicine	Saar, 1991
Russian Federation, far east	Medicinal mushrooms in nature	Bulakh, 2001
Tanzania (United Republic of)	Compares use of wild edible fungi with customs in Hunan in China	Härkönen, 2002; Härkönen, Niemelä and Mwasumbi, 2003
Turkey	Edible fungi, part of an ethnobotanical study	Ertrug, 2000
Zambia	Customs and folklore about mostly edible species	Piearce, 1981

Note: See also Volume 3 (1–2) of the *International Journal of Medicinal Mushrooms* for abstracts from a conference on medicinal mushrooms, many of which have a ethnoscientific slant.

forestry business does not mean that collecting wild edible fungi is either an attractive or economic proposition, even to people desperate for work (Tedder, Mitchell and Farran, 2002).

NUTRITION AND HEALTH BENEFITS

Useful macrofungi consist of those with edible and medicinal properties[9]. There is no easy distinction between the two categories. Many of the common edible species have therapeutic properties; several medicinal mushrooms are also eaten (Table 14). *Ganoderma* species (*ling zhi* or *reishi*) are the most valuable medicinal mushrooms (Plate 9): the global value of ganoderma-based dietary supplements has been estimated to be US$1.6 billion (Chang and Buswell, 1999).

Lentinula edodes and *Volvariella volvacea* are widely cultivated edible fungi with medicinal properties. Only *Inonotus obliquus*, out of the 25 medicinal species listed in Table 14, appears not to be cultivated. Of the 182 medicinal fungi reported in Annex 3 only 5 percent are ectomycorrhizal (see Reshetnikov, Wasser and Tan, 2001). This is probably an underestimate (Mao, 2000) since research efforts have concentrated on saprobic species that can be cultivated, thus providing a guaranteed supply and uniformity of product.

There has been a spectacular increase of interest and commercial activity concerned with dietary supplements, functional foods and other products that are "more than

[9] Ceremonial, religious and other non-practical uses of wild fungi are of relative minor importance and are not discussed here (see Davis, 1996 and Riedlinger, 1990, for more information).

TABLE 14
Nutritional composition of some wild edible fungi

		Composition, percentage dry weight			
Binomial	Country	Protein	Carbohydrate	Fat	Mineral matter (ash)
Amanita caesarea	France? (1)	15	nk	14	10
Amanita loosii	Democratic Republic of the Congo (2)	20	nk	nk	nk
Amanita rubescens	Mexico (3)	18	nk	nk	nk
Boletus edulis	Turkey (7)	38	47	9	1
Boletus edulis	Finland (8)	23	nk	2	7
Boletus erythropus	Jordan (5)	15	57	1	8
Boletus frostii	Mexico (3)	16	nk	nk	nk
Boletus loyo	Chile (12)	22	50	1	6
Cantharellus cibarius	Turkey (6)	21	62	5	2
Cantharellus cibarius	Democratic Republic of the Congo (10)	15	64	5	13
Lactarius phlebophyllum	United Republic of Tanzania (7)	30	51	9	5
Lactarius deliciosus	France? (1)	23	nk	7	6
Lactarius deliciosus	Chile (4)	27	28	7	6
Lactarius indigo	Mexico (3)	13	nk	nk	nk
Lactarius torminosus	Finland (8)	21	nk	2	7
Lactarius piperatus	Turkey (6)	27	65	2	1
Ramaria flava	Mexico (3)	14	nk	nk	nk
Ramaria flava	Finland (8)	24	nk	2	6
Russula cyanoxantha	France? (1)	17	nk	8	8
Russula delica	India (9)	17	nk	nk	nk
Russula sp.	Democratic Republic of the Congo (10)	29	55	6	6
Suillus luteus	Chile (4)	20	57	4	6
Suillus granulatus	Chile (4)	14	70	2	6
Terfezia claveryi	Iraq (11)	8	17	nk	10
Termitomyces microcarpus 1	United Republic of Tanzania (7)	49	29	10	11
Termitomyces microcarpus 2	United Republic of Tanzania (7)	35	37	6	23
Termitomyces microcarpus	Democratic Republic of the Congo (10)	33	38	5	14
Tricholoma populinum	Canada (13)	13	70	9	7
Tricholoma saponaceum	France? (1)	5	nk	7	8
Tirmania nivea	Iraq (11)	14	21	nk	5

nk – not known. Figures rounded to nearest whole number.
Sources: (1) Kiger, 1959 – assumed to have tested specimens from France but not stated; (2) Degreef *et al.*, 1997; (3) Leon-Guzman, Silva and Lopez, 1997; (4) FAO, 1998b; (5) Ereifej and Al-Raddad, 2000; (6) Caglarirmak, Unal and Otles., 2002; (7) Härkönen, Saarimäki and Mwasumbi, 1994a; (8) Kreula, Saarivirta and Karando, 1976; (9) Purkayastha and Chandra, 1985; (10) Parent and Thoen, 1977; (11) Al-Naama, Ewaze and Nema, 1988; (12) Schmeda-Hirschmann *et al.*, 1999b; (13) Turner, Kuhnlein and Egger, 1987.

just food" (Etkin and Johns, 1998; Wasser *et al.*, 2000). Although these new products have clear economic potential, their relevance to developing countries is at present still marginal. Medicinal wild fungi are collected in China. There is a substantial trade of *Cordyceps sinensis* in Sichuan (Plate 9) (Priest, 2002, personal communication: *Edible and medicinal fungi in China and general information*; Winkler, 2002) and in other countries such as Nepal. Rural people earn substantial amounts from commercial harvesting.

The main benefits of wild useful fungi are, however, as food. They are collected, consumed and sold in over 85 countries (Annexes 1 and 2) and their contribution to diets is discussed below.

Nutritional value

The constituents of an edible fungus are not necessarily a good guide to nutritional value (Breene, 1990). The digestibility of different components varies, while analytical methods are not always reliably used in testing (Crisan and Sands, 1978; Lau, 1982). The use of different techniques for analysing nutritional value also limits a comparison of results from different studies (Degreef *et al.*, 1997). Estimates of (usable) protein content should exclude chitin present in fungal cell walls, for example. This is not always observed in studies.

A summary of nutritional analyses is presented in Table 14. Note the good protein and mineral content of key wild edible species in their dry state. (Moisture content varies between about 85 and 95 percnet for the fleshy mushrooms and similar types.) Edible species are low in fat, contain essential amino acids and useful minerals and, though they are not energy-providing foods (Table 16), they are a substantially better source of nutrition than is often assumed or inferred (Richards, 1939).

Contribution to diet

Tables 15 and 16 compare the nutritional value of edible fungi with other foodstuffs. These data confirm that wild edible fungi are nutritious and a suitable alternative for well-known foodstuffs. They compare favourably using standard measures that assess the nutritional value of foods. The contribution to diet will depend on the amounts eaten by people, the species involved and the frequency of consumption (see below).

People regularly eat wild edible fungi in many countries and they make a valuable and often essential contribution to diets, as shown by a study in Malawi (Abbott, 1999). This detailed study of eating habits in villages revealed that 1.3 kg of dried leafy vegetables and/or wild edible fungi was enough (when rehydrated) to feed a family of four for two weeks (Abbott, 1999).

BOX 7

Amino acids, protein and the nutritional value of wild edible fungi

Various measures (scores, indexes, values) based on amino acid composition are used to compare the nutritional value of wild edible fungi with other foods. Fat and carbohydrate content are of less interest because they are rarely limiting factors in diets. Feeding studies of edible fungi would provide the most direct evidence of nutritional value but until now have not been carried out.

The Amino Acid Score is based on the amount of the most limiting amino acid present in a food item in comparison with a reference protein (e.g. hens' eggs). The Essential Amino Acid Index measures the presence of amino acids that people cannot synthesize and gives a stronger indication of potential nutritive value. However, this index does not indicate how well these essential amino acids are retained and used by the body, which is the reason for computing the Biological Value, itself derived from the Essential Amino Acid Index.

The Essential Amino Acid Indexes for wild edible fungi compare favourably with other foods (Table 16). Given that there are restricted sources of protein for rural people in the developing countries, the contribution of wild edible fungi is more important than widely recognized. The Nutritional Index allows comparisons to be made between wild edible fungi with small amounts of high quality protein and those that have large amounts of a lower nutritional value. The data in Table 15 show the greatest range of values for the limited number of species tested.

The ultimate contribution made by wild edible fungi to diets depends not only on their intrinsic value as calculated by these measures, but the amounts (and species) eaten in comparison with other foods. The nutritional analyses show that wild edible fungi are a valuable source of protein in the developing counries and have the potential to contribute more to human diets in many countries.

After Crisan and Sands (1978).

TABLE 15
Estimated nutritional values of some edible fungi

Species	Essential Amino Acid Index	Biological Value	Amino Acid Score	Nutritional Index
Agaricus bisporus *	86.8	83.0	65.0	22.0
Cantharellus cibarius	94.2	91.0	68.0	3.31
Macrolepiota procera	98.7	95.9	90.0	7.4
Suillus granulatus	89.7	86.1	73.6	13.5
Termitomyces spp.	86.3	82.4	–	23.9
World species	87.6	83.8	61.6	16.0

* cultivated. Based on FAO reference patterns and mean values for species from several sources. Unpublished data prepared by Graham Piearce. See Box 6 for a discussion of nutritional indicators.

TABLE 16
A general comparison of nutritional values of various foods compared to mushrooms

Essential amino acid index	M
100 Pork, beef, chicken	
99 Milk	
91 Potatoes, beans	
88 Maize	
86 Cucumbers	
79 Groundnuts	
76 Spinach, soybeans	
72 Cabbage	
69 Turnips	
53 Carrots	
44 Tomatoes	

Amino acid score	M
100 Pork	
98 Beef, chicken	
91 Milk	
63 Cabbage	
59 Potatoes	
53 Groundnuts	
50 Maize	
46 Beans	
42 Cucumbers	
33 Turnips	
31 Carrots	
28 Spinach	
23 Soybeans	
18 Tomatoes	

Nutritional index	M
59 Chicken	
43 Beef	
35 Pork	
31 Soybeans	
26 Spinach	
25 Milk	
21 Beans	
20 Groundnuts	
17 Cabbage	
14 Cucumbers	
11 Maize	
10 Turnips	
9 Potatoes	
8 Tomatoes	
6 Carrots	

M – shaded column shows the range of values for mushrooms. Indexes and scores calculated against reference patterns published by FAO; *biological values* closely follow *essential amino acid indexes*. Data after Crisan and Sands (1978).

The shelf-life of wild edible fungi can be short but harvests are also preserved in a number of ways. In the Russian Federation and China wild edible fungi are commonly preserved in brine (Plate 8). Russians also freeze wild edible fungi for later use (Vladyshevskiy, Laletin and Vladyshevskiy, 2000). In southern Africa, edible fungi are eaten fresh and less commonly dried. Throughout the miombo region of southern Africa wild edible fungi are an important source of nutrition at a time of year when other food supplies are low – the so-called "famine months". Here the normal diet consists of *nsima* (a maize or cassava-based porridge) to which relishes are added (Plate 6). The relishes provide key nutrients and add piquancy to the bland *nsima*.

Information on the amounts of wild edible fungi consumed includes:

- **Mozambique**: in the north, close to the border with Malawi, people collect from 6 to 10 kilograms of wild edible fungi during a season (December to March). It was estimated that each household ate 72 to 160 kg per year. Average consumption of *Termitomyces schimperi* was reckoned to be 30–35 kg per household per year. Similar eating habits might be reasonably expected to occur in Malawi and other miombo regions. (Masuka in Boa *et al.*, 2000).
- **Zimbabwe**: households eat up to 20 kg in a productive year but only 5–10 kg in deforested areas (Masuka, 2002, personal communication: *Collection of mushrooms in Zinbabwe*).
- **Russian Federation – Siberia**: people collect 15–100 kg in a year and eat 80–90 percent directly. The population of Krasnoyarsk region is three million over an area of 2.3 million km^2; it is estimated that 40 percent of families collect wild edible fungi, for personal use, recreation or sale (based on interviews with 500

respondents). Use of wild edible fungi has increased by 200–300 percent in recent years and now provides 30–40 percent of household income. (Vladyshevskiy, Laletin and Vladyshevskiy, 2000).

As a general rule, the poorer the people the more likely they are to collect and use wild edible fungi. Some traditions are lost as people become better educated and live away from the land and they show an increasing reluctance to eat all but the most common species (Box 3) (Lowy, 1974). In the Republic of Korea, China, the Russian Federation and Japan the tradition of eating wild edible fungi is much stronger and appears to have withstood the changes experienced elsewhere.

Rural people eat wild edible fungi both as a matter of choice and as a food of last resort. Little reliable information is available, however, on the use of wild fungi as famine foods. In the Russian Federation, food distribution systems have collapsed and state subsidies for food have disappeared, forcing people "back to the land". A renewed dependency on natural products has developed and traditions of collecting and eating wild edible fungi have been reinforced. The extent of these changes is not well understood but emphasizes again that closeness to the land is associated with eating wild edible fungi.

Contribution to health

Medicinal fungi are routinely used in traditional Chinese medicine (TCM) and awareness of their uses is increasing (Ying *et al.*, 1987; Hobbs, 1995). Wild medicinal fungi are also collected and used in Mexico and several other countries (Table 13) but widespread and regular use is most closely associated with China and Asian people. Medicinal fungi are often sold in Chinese markets though the contribution from wild harvests is still unclear (Chamberlain, 1996).

Worldwide, the majority of sales are from cultivated sources though many species are also collected from the wild (Table 17). The incentive for collecting wild *Cordyceps sinensis* in Tibet Autonomous Region, Sichuan (Winkler, 2002) and other parts of China (see distribution map in Mao, 2000) is to earn money (Plate 9). Beyond China there is no discernable international trade in medicinal fungi.

The therapeutic benefits of wild fungi are summarized below (Table 17), noting that many are also consumed as food.

LOCAL MARKETING AND INCOME

There are two distinct patterns of wild edible fungi use: for subsistence or personal use and commercial harvesting. Information about personal collections is scarce, but the extent of this practice is global and there are increasing reports that help to demonstrate the importance of WEF to rural people in developing countries. Many more species are eaten locally compared to the small number involved in commercial harvesting.

Finland has the most detailed information on personal collections of wild edible fungi. Wild edible fungi are a less important part of the diet in Finland today, in times of relative affluence, but there is still government support for collecting them. There is a stronger tradition of collecting and consuming wild edible fungi in the east of Finland, a region where Karelian people originally from the Russian Federation have settled. Around 25 percent of Karelian families collect to sell in markets, though the amounts vary from year to year because of fluctuating harvests. 1976 was a poor year and about 45 percent of families interviewed did not collect any wild edible fungi during this period. Poorer communities collected more often to sell in local markets (Härkönen, 1998).

The total amounts sold in local markets can be considerable (Table 18). Anecdotal evidence from China points to huge quantities collected and taken to markets in small towns and from there to larger cities (Plate 9). Preserving wild edible fungi in brine is an important feature of this trade and it allows much larger quantities to be offered for sale. The financial contributions to rural livelihoods are not known though the

TABLE 17
Properties and features of 25 major medicinal macrofungi

BINOMIAL	MEDICINAL PROPERTIES	USED AS FOOD?	WILD COLLECTION[1]	CULTIVATED	COMMERCIAL PRODUCT
Agaricus blazei	1	"edible"	+	yes	no
Agrocybe aegerita	4	yes	+	yes	yes
Armillaria mellea	4	yes	++	yes	yes
Auricularia auricula-judae	5	yes	++	yes	yes
Dendropolyporus umbellatus	4	no	+	yes	no
Flammulina velutipes	5	yes	++	yes	yes
Fomes fomentarius	2	no	+	yes	yes
Ganoderma applanatum	4	no	+	yes	yes
Ganoderma lucidum	11	"edible"	+	yes	no
Grifola frondosa	7	yes	+	yes	yes
Hericium erinaceus	4	yes	+	yes	yes
Hypsizygus marmoreus	1	yes	+	yes	no
Inonotus obliquus	4	no	++	no	no
Laetiporus sulphureus	2	yes	++	yes	yes
Lentinula edodes	11	yes	+	yes	no
Lenzites betulina	2	no	?	?no	yes
Marasmius androsaceus	2	?yes	?	?yes	no
Oudemansiella mucida	1	"edible"	++	yes	no
Piptoporus betulinus	2	no	++	yes	yes
Pleurotus ostreatus	5	yes	+	yes	yes
Pleurotus pulmonarius	3	yes	+	yes	yes
Schizophyllum commune	5	yes	++	yes	no
Trametes versicolor	5	"edible"	+	yes	no
Tremella fuciformis	5	"edible"	+	yes	yes
Volvariella volvacea	4	yes	+	yes	yes

[1] + minor importance; ++ significant amounts collected. Both assessments are in relation to the total amounts used globally, including cultivated production.
Note: The 14 different medicinal properties consist of: 1 – *Antibiotic* (includes antifungal, antibacterial, antiparasitic **but not** antiviral); 2 – *Anti-inflammatory*; 3 – *Antitumour*; 4 – *Antiviral*; 5 – *Blood pressure regulation*; 6 – *Cardiovascular disorders*; 7 – *Hypercholesterola emia, hyperlipidaemia* [high cholesterol, high fats]; 8 – *Antidiabetic*; 9 – *Immune-modulating*; 10 – *Kidney tonic*; 11 – *Hepatoprotective*; 12 – *Nerve tonic* (? antidepressant; vague); 13 – *Sexual potentiator*; 14 – *Chronic bronchitis* (against).
Source: Wasser and Weis, 1999a.

widespread sale of wild edible fungi within China and the substantial export business (over 60 percent of *Boletus edulis* imported by Italy comes from China – Borghi [2002, personal communication: *Porcini and other commercial wild edible fungi in Italy*]) clearly demonstrates that substantial amounts of money are earned.

Experiences in Malawi showed that money earned by local collectors is small but substantial, and that there is an expanding local market for wild edible fungi (Boa *et al.*, 2000). Women frequently go on collecting trips in many parts of southern Africa and a number of reports confirm the importance of this activity during the three- to four-month season each year (Richards, 1939; Thomson, 1954).

The distance from collecting sites to potential markets is a crucial factor in selling wild edible fungi. The roadside markets at Liwonde in Malawi are close to the forest areas where wild edible fungi are collected. The road is the main thoroughfare from Blantyre to Lilongwe and the makeshift stalls sell round 5 tonnes of wild edible fungi during a four-month season. There is no shortage of people wanting to collect and sell, and this has led to increased competition for fungal resources: people now have to walk further to collect (Lowore and Boa, 2001).

The market structure in Malawi is typical of many African countries (e.g. Sierra Leone: Down, 2002, personal communication: *Wild edible fungi Sierra Leone*): small-scale and local. Sales at Liwonde and elsewhere depend on the flow of traffic and some days few buyers stop. Some traders wait until the end of the day and buy the unsold

BOX 8
Permits and regulating the collectors

One of the inevitable consequences of commercial harvesting is the introduction of permits. From Bhutan to Serbia these are ostensibly introduced to regulate the impact of collectors and collecting on future production of wild edible fungi, yet there is little evidence that the money paid to local authorities is invested in the resources needed to police activities.

In Castilla León, northwest Spain, the permit system for collecting *Lactarius deliciosus* collapsed in Buenavista de Valdavia when only four people bought permits in 2002, at a cost of US$30 for a six-week season. The other collectors decided this was no longer necessary, mainly because the guards from the Servicio de Protección de la Naturaleza proved to be increasingly ineffective in checking permits. Local collectors were concerned about the influx of outsiders to collect the *níscalos* and were insulted when asked to show their permits. There is no obvious friction between the local people and visiting collectors from nearby villages, but several people said the permit system should be reinstated since they were worried about the long-term prospects for mushroom production.

Around Borgo Val de Taro, Parma, in northern Italy, the permit system appears to work more effectively. The local authority publishes the regulations each year, stating the conditions and costs of collecting WEF. The rates vary from around US$5 for a one-day permit for local residents with slight increases for non-residents. The differences are more marked for the six-month permits, with non-residents paying up to twice as much (up to US$100) as local people. Collecting is restricted to three or four days a week and a daily harvest of between 3 and 5 kg. This area is noticeably better off than Buenavista de Valdavia, where the need to earn money from *níscalos* is more urgent.

In France, the increase in people collecting wild edible fungi has prompted the introduction of more formal rules regarding when and how much can be collected. Daily limits of 5 kg are stated with no collecting allowed on Tuesdays and Thursdays. A yearly permit costs around US$120.

Sources: Spain – de Román (2002, personal communication: *Trade in níscalos from North Spain to Catalonia and truffle production*), *Italy* – author's observations and Zambonelli (2002, personal communication: *Truffles, and collecting porcini in Italy*); *France* – Bérelle (2002).

produce, moving it quickly to more central markets in the bigger cities. The prices they offer are low but the alternatives are either to dry the fungi or discard them. Local markets in Madhya Pradesh, India, are also small-scale (Harsh, Rai and Soni, 1999) and appear to operate in a similar manner, but within towns rather than by the roads.

In the Russian Federation the collapse of state organizations and state buying has significantly affected the amounts of money people can earn from wild edible fungi (Table 18). Previous displeasure about the low prices offered by the state are, in hindsight, viewed less harshly following the collapse of local markets (Vladyshevskiy, Laletin and Vladyshevskiy, 2000).

The removal of state control in China has unleashed a greater entrepreneurship, though it has not been without its failures. Factories for processing *matsutake* in Sichuan are barely surviving (Winkler, 2002); similar facilities for producing *ganbajum* (*Thelephora ganbajum*) never operated effectively and were eventually shut down (Rijsoort and Pikun, 2000). The local trade in *ganbajum* has continued, though collectors spend longer in cleaning their harvest for market (up to two hours per kilogram). Consumers pay a higher price for better quality produce.

NATIONAL AND INTERNATIONAL TRADE

The international trade in wild edible fungi has taken place for many years. In the 1880s New Zealand exported ear fungus (*Auricularia polytricha*) to China (Colenso, 1884–85; Hall, Zambonelli and Primavera, 1998). In 1868, France exported a staggering 1 500 tonnes of truffles (*Tuber* spp.) to Italy (Ainsworth, 1976). Italy has long imported *Boletus edulis* and truffles from different countries (Plates 4 and 5):

TABLE 18
Local collection, marketing and use of wild edible fungi

Country	Collections and use	Amount	Source
Bhutan	People regularly collect for personal consumption and sell in markets. Some *matsutake* were sold previously in markets but mostly by accident. People sell to agents who sell to exporters.	not known for personal collections	Namgyel, 2000
Chile	*Cyttaria* spp., total collection in one season, for local sale and consumption.	500–700 kg	Schmeda-Hirschmann *et al.*, 1999a
China (Sichuan)	Many species collected and eaten. *Matsutake* "discovered" by Japanese in 1988. Exported through Kunming and ?Chengdu. *Matsutake* are bought by traders with access to suitable transport, taken to a town 65 km away and sold on at a profit of 75%.	not known for personal collections	Winkler, 2002; Yeh, 2000
China (Yunnan)	Daily collection of edible species in Guilong, Deqing over an eight-month season. Sold locally.	60–100 kg	Rijsoort and Pikun, 2000
Congo (Democratic Republic of the)	Annual consumption in Shaba region from local collection.	20 000 tonnes	Degreef *et al.*, 1997
Estonia	Self-picked mushrooms, average annual amount per capita	2.4 kg	Paal and Saastamoinen, 1998
Finland	1. *Gyromitra esculenta* bought by trade in (a) 1988, (b) 1996. The Russian Federation is another possible source. 2. About two million people involved in collecting WEF and berries for personal use or for sale. An average of 8% of collectors sold their harvest in 12 districts, 25% in North Karelia and not at all for two districts (1976 survey). Export activity limited.	(a) 109 tonnes (b) 26 tonnes	1. Härkönen, 1998 2. Pekkarinen and Maliranta, 1978
Germany (Munich)	For sale during summer of 1902, all species. Source(s) of wild edible fungi not known.	400 tonnes	Arnolds, 1995
India (Himalaya)	Daily harvest of morels by experienced collectors, all for export.	Up to 1 kg	Singh and Rawat, 2000
India (Madhya Pradesh)	*Termitomyces heimii* sold in 15 markets during one year for local consumption. Cannot be stored for more than a day; some are dried and eaten later. *T. heimii* does not get price premium it deserves. Medicinal polypores are collected but bought at low prices compared to retail price in New Delhi.	2.5 tonnes	Harsh, Rai and Soni, 1999
Italy	*Tuber* spp. collected in average year, including 50% hike for black market activity. Sold locally.	160 tonnes	Hall *et al.*, 1998a
Malawi (Liwonde)	All edible species, sold in 2000 over two months, from approx. 10 small stalls.	5 tonnes	Boa *et al.*, 2000
Mexico (Mexico City)	*Huitlacoche* (maize infected with *Ustilago maydis*) sold in markets	300–400 tonnes	Villanueva, 1997
Mexico (Tlaxcala)	Harvest from one day's collecting, all species	4–5 kg	Montoya-Esquivel *et al.*, 2001
Russian Federation (central Siberia)	Individual collection of all species in favourable years. 80–90% are for personal consumption, the rest are sold. More families are freezing harvests. In north Taiga people eat WEF almost every day. Marketing has collapsed as state organizations have declined: previously GOSPROMKHOV bought up to 1 000 tonnes at fixed prices when harvest was good and purchase prices were lower.	15–100 kg	Vladyshevskiy, Laletin and Vladyshevskiy, 2000
Tanzania (United Republic of)	Sold by the road (often close to the spot where *Termitomyces* grow) and in markets. There are no known exports from the United Republic of Tanzania.	not known	Härkönen, 2002
Turkey	Collections from 13 villages of (a) *Cantharellus cibarius*; (b) *Boletus edulis*; (c) *Morchella* sp.; (d) *Lactarius* sp.– total value of US$107 000. Most for local sale. Total volume 26 tonnes. Data for 1990.	(a) 7.6 tonnes (b) 2.5 tonnes (c) 2.3 tonnes (d) 11.1 tonnes	Cavalcaselle, 1997
Zimbabwe	Collection of *Boletus edulis* per person per day, for export only.	15–20 kg	Masuka, 2002, pers. comm.: *Collection of mushrooms in Zimbabwe*

Note: Amounts are fresh weight or presumed to be so in the absence of other information.

the former Yugoslavia began exports of *B. edulis* in the 1970s (Borghi, 2002, personal communication: *Porcini and other commercial wild edible fungi in Italy*).

The exports of *matsutake*, chanterelles, morels and other "exotic" wild edible fungi are a more recent event, and where France once exported truffles to Italy, China now exports *Tuber sinosum*. The last 20 or 30 years has seen an increasing movement of chanterelles, morels and *Boletus edulis* from the southern to the northern hemisphere. Within Europe, the local supply of wild edible fungi has failed to meet an expanding demand for "exotic mushrooms" (Plate 9).

The increased demand has provided commercial opportunities for countries in eastern Europe, Turkey, and Mexico – to name a few. The United States and Canada have increased exports of a number of wild edible fungi, though they are most associated with *matsutake* sent to Japan (Box 4). The Japanese demand for *matsutake* has had an important effect on the livelihoods of people in Asia and North America. Tables 21, 22 and 23 provide an overview of the global trade in *matsutake*.

The price paid for *matsutake* varies considerably, depending on annual harvests around Asia and in the United States and Canada. The financial benefits to collectors are difficult to quantify, although the signs of increased wealth are clear to see in parts of Sichuan. In Kyanbga the money earned from selling *matsutake* and *Cordyceps* spp. provides 60 percent of cash income (Winkler, 2002). The enthusiasm for collecting, clandestine planning of trips (rising early in the morning and hunting with torches in Bhutan: Namgyel 2000) and sometimes violent clashes between collectors (Yeh, 2000) indicates the perceived attraction of the potential financial rewards.

The quality of *matsutake* significantly affects prices obtained by collectors. Exports from the Republic of Korea are worth a similar amount to the Democratic People's Republic of Korea when averaged over a five-year period (Table 23) even though the average volume exported over the same period was only about 25 percent of that for the Democratic People's Republic of Korea. The Italian traders have provided technical support to improve and maintain the quality of *Boletus edulis* exports from Serbia, and there has been a steady increase in the amounts of money earned at a national level (Borghi, 2002, personal communication: *Porcini and other commercial wild edible fungi in Italy*).

The amounts paid per kilogram for truffles (*Tuber* spp.) and *matsutake* generate much interest but this is not necessarily reflected in the amounts earned by collectors. It is possible to make a good living from truffle collecting but the numbers who benefit are relatively small (Plate 4). Rural people earn useful amounts in a short period of time from collecting morels (*Morchella* spp.) in India (Prasad *et al.*, 2002) and Pakistan (*Pakistan Economist*, 2001), but trade in Nepal and Afghanistan appears to be less lucrative. The morels are collected in the Himalaya and collectors can earn US$ 6–7 per day. The total money earned in a season provides 20–30 percent of the annual cash income in 140 villages (Singh and Rawat, 2000) and an annual income of US$150 from another survey of 1 600 families in 40 villages (Prasad *et al.*, 2002)

In Turkey, around 11 tonnes of fresh *Lactarius delicious* were sold in 13 villages (Table 18). The total annual value of four key wild edible species was around US$100 000, a substantial source of local income. The role of traders is important in facilitating local markets and the international trade. They provide transport, credit and even technical support. More importantly, they provide some guarantee of a sale. They also benefit financially from the higher prices when produce is sold on, and this has attracted some criticism (Harsh, Rai and Ayachi, 1993). But without traders there would be no export markets and this would reduce the substantial benefits earned locally and nationally from the commercial harvesting of wild edible fungi.

The sale of harvesting permits (Chapter 3, section *Regulating collection*) and local taxes are other sources of potential revenue. It has been estimated that twice the officially recorded harvests of *Tuber* spp. take place in a year (Hall, Zambonelli and Primavera, 1998). Similar estimates and higher have been made for former Yugoslavia

TABLE 19
World production of cultivated mushrooms

ITEM	1986	1989/90	1994	1997	2001*
World production (tonnes)	2 182 000	3 763 000	4 909 000	6 202 000	7 500 000
China production (%)			54	70	
Value world production (US$ billion)		7.5	16		22.5
Agaricus bisporus (%)	56	38	38	32	nd
Lentinula edodes (*shi'itake*) (%)	14	10	17	25	nd
Pleurotus spp. (%)	8	24	16	14	nd

* 2001 figures are estimates based on 5 percent annual increase in volume and 5 percent increase in value at 1994 prices.
Sources: Chang, 1991; Chang and Miles, 1991.

TABLE 20
Value of wild useful fungi collected by country of origin

COUNTRY	COLLECTION AND EXPORT	VALUE US$ (MILLIONS)	SOURCE
Canada	Before tax revenue of 16 companies involved in harvesting, buying or selling all wild edible fungi. Around 6 000 collectors are involved. Range is for "bad" and "good" years.	15–27	Wills and Lipsey, 1999
China (Sichuan)	(a) *Cordyceps* annual harvest 1949 to mid-1980s. (b) *Cordyceps sinensis* harvest in Litang	(a) 5–20 (b) 1.2–1.8	Winkler, 2002
China (west Sichuan)	*Tricholoma matsutake*, income for farmers.	5–6	Winkler, 2002
Chile	Salted (*salmuerados*) and dried (*deshidratados*) wild edible fungi exported, 1980 – 1990. Annual value: (a) average (b) range	(a) 1.8 (b) 1.3–2.8	FAO, 1993a
Mexico (in six states)	*Tricholoma magnivelare* for export: (a) 1996; (b) 1997. Involves 3 000 families.	(a) 1.1 (b) 0.6	www.semarnat.gob.mx
Turkey	*Terfezia boudieri*, *Boletus* sp., *Morchella* sp., *Cantharellus cibarius* for export in (a) 1991 (b) 1999	(a) 14.4 (b) 9.5	Sabra and Walter, 2001
United States	(a) morels; (b) chanterelles; (c) *matsutake*; (d) boletes. Data for 1992.	(a) 5.2 (b) 3.7 (c) 8 (d) 2.3	Schlosser and Blatner, 1995
Zimbabwe	*Boletus edulis* for export in one year. Said to involve 2 000–5 000 collectors.	1.5	Boa *et al*., 2000

and a range of commercially important species (Ivancevic, 1997). Revenue from permits and taxes does not always reflect the amounts of wild edible fungi collected.

The income from commercial harvesting is uncertain. Fluctuating harvests and competing supplies from other countries can result in wide fluctuations in prices offered, particularly with truffles and *matsutake*. The quality of the collected produce is also important and attention to this detail is a simple way of maximizing income for collectors. The increased supply of chanterelles to the United Kingdom during the 1990s has depressed the wholesale price by two-thirds (Livesey, 2002, personal communication: *Import of wild edible fungi to the UK*), though increased volumes exported by Poland (Table 20) have increased total revenues.

The overall effect is that there are few who make their sole living from collecting wild edible fungi. There is no evidence from commercial collecting (Dyke and Newton, 1999) to support a quoted income of around US$3 000 from a week's endeavours in the United Kingdom (Rotheroe, 1998). The commercial trade in wild edible fungi has, however, earned many countries substantial amounts of money. The Democratic People's Republic of Korea earned US$150 million from *matsutake* exports to Japan over a five-year period (Table 23). More detailed studies are needed to examine how collectors benefit from this trade.

The patchy data on volumes of exports for key commercial species suggest that relatively small amounts are involved (Table 24). Poland exported just over 9 000 tonnes of chanterelles in 1984, the former Soviet Union around 3 000 tonnes. Turkey exported 730 tonnes of *Boletus edulis* in 1990 while India, Pakistan, Nepal, Afghanistan and possibly Iran collect around 2 000 tonnes fresh weight of morels in a year. The benefits to rural livelihoods are significant and widespread and large numbers of people earn significant amounts of money.

World trade in cultivated mushrooms

There has been a spectacular increase in world production over the last ten years (Table 19). In 1997 *shi'itake* (*Lentinula edodes*) and *Pleurotus* spp. together exceeded the value of sales of *Agaricus bisporus*, a mushroom celebrated more for its shape than its taste. An estimate of world production for 2001, based on figures for 1997, puts the

TABLE 21
***Matsutake* 1: domestic production and imports in tonnes to Japan, 1950–99**

Year	Domestic prod.	Imports	% import	Domestic and imports	Consumption as a % of 1950
1950	6 448	0	0	6 448	
1955	3 569	0	0	3 139	49
1960	3 509	0	0	3 509	54
1965	1 291	0	0	1 291	20
1970	1 974	0	0	1 974	31
1975	774	0	0	774	12
1980	457	362*	44	819	13
1982	484	551	53	1 035	16
1984	180	1 082	86	1 262	20
1986	199	980	83	1 179	18
1988	406	1 430	78	1 836	28
1989/90	199	2 210	92	2 409	37
1993	na	1 943	–	[1 943]	
1994	na	3 622	–	[3 622]	
1995	na	3 515	–	[3 515]	
1996	na	2 703	–	[2 703]	
1997	na	3 059	–	[3 059]	
1998	257	3 248	93	3 505	54
1999	147	2 674	95	2 821	44

* first year that imports are noted. na – data not available. Domestic production from 1993 to 1997 thought to be around 200 tonnes per year.
Source: Data have been collected from various authors. The original source appears to be Japanese trade statistics. See www.fintrac.com for data from 1993 to 1997.

TABLE 22
***Matsutake* 2: exports to Japan in tonnes by various countries, 1993–97**

Country	1993	1994	1995	1996	1997	Average tonnes/year	Five-year value US$ millions
Bhutan*	1	1	2	3	3	**2**	1
Canada**	279	447	340	510	618	**439**	95
China*	1 064	1 127	1 192	1 152	1 076	**1 122**	270
Korea (Democratic People's Republic of)*	383	1 760	1 141	541	615	**888**	156
Korea (Republic of)*	131	139	633	170	249	**264**	169
Mexico**	2	22	36	23	9	**18**	6
see below	*(26)*	*(35)*	*(56)*	*(42)*	*(14)*		
Morocco***	20	73	1	86	125	**61**	12
Turkey***	0	2	4	44	80	**26**	4
United States**	51	47	164	172	284	**144**	33

* *Tricholoma matsutake*. ** *T. magnivelare*. *** probably *T. caligatum*. Includes fresh and chilled.
Note: The export tonnage from a "Mexican Government database" (Martínez-Carrera *et al*., 2002) is shown in italics and includes data for 1998 (24 tonnes); 1999 (14 tonnes) and 2000 (4 tonnes).
Source: www.fintrac.com.

TABLE 23
Matsutake 3: value of exports to Japan by various countries, 1993–97

Country	1993 Yen, million	1994 Yen, million	1995 Yen, million	1996 Yen, million	1997 Yen, million	Total Yen, million	Total US$, million
Bhutan	5	4	9	17	16	**51**	**0.5**
Canada	1 840	1 891	1 506	2 690	2 559	**10 486**	**95**
China	5 494	5 746	5 249	6 631	6 579	**29 699**	**270**
Korea (Democratic People's Republic of)	2 291	6 928	4 074	1 060	2 794	**17 147**	**156**
Korea (Republic of)	2 321	2 653	6 719	3 076	3 815	**18 584**	**169**
Mexico	78	100	206	156	73	**613**	**6**
Morocco	117	340	6	368	449	**1 280**	**12**
Turkey	0	4	12	140	256	**412**	**4**
United States	491	253	782	931	1 153	**3 610**	**33**
Total	*12 637*	*17 919*	*18 563*	*15 069*	*17 694*	**81 882**	**745**
Grand total *(US$, million)*	115	163	169	137	161		745

US$1 = 110 Yen. Grand total includes several countries that were minor and irregular exporters. Data include fresh and chilled *matsutake*.
Source: www.fintrac.com.

TABLE 24
Volume of exports of named wild edible fungi from selected countries (in tonnes)

Country	Year	*Boletus edulis*	Chanterelles	Morels*
Baltic states (86% Lithuania)	1998	nd	3 500	nd
India	annual	none?	nd	50-60
Pakistan	1999	none	none	79
Poland	1984	nd	9 179	nd
South Africa	annual	100–200	none	none
Turkey	1989	22	11	47
	1990	730	160	nd
	1996	nd	13	152
	1997	nd	18	100
	1998	nd	375	46
	1999	nd	94	104
	2000	nd	15	44
Yugoslavia (former – now Serbia and Montenegro)	1993	5 186	2 605	37
	1994	1 212	631	2
	1995	3 792	1 502	3
Zimbabwe	annual	100	20-30	none

nd – no data. none – no evidence of exports. * dry weight. All other data are assumed to be fresh weight.
Sources: *Pakistan Economist*, 2001; Boa *et al*., 2000; Gurer, 2002, personal communication: *Unpublished trade data on wild edible fungi for Turkey*; Kaul, 1993; Kroeger, 1985; Pott, 2002, personal communication: *Export of* Boletus edulis *from South Africa*; Sabra and Walter, 2001

global value of cultivated mushrooms at around US$23 billion. This exceeds the value of many other commodities.

The trade in wild edible fungi and the business of cultivated mushrooms have both steadily expanded. Packets of wild and cultivated species are sold in shops (Plate 9). Sales of wild edible fungi have risen steadily as the range of commercial species on sale in the United Kingdom has increased. In China, customers have been observed to prefer the wild species, when in season, to the cultivated mushrooms that are available all year round (Priest, 2002, personal communication: *Edible and medicinal fungi in China and general information*).

Cultivated mushrooms are now China's biggest "vegetable" export and there are significant numbers of relatively small-scale producers in countries such as Viet Nam and Indonesia (Gunawan, 2000). Both China and Viet Nam export cultivated mushrooms to Europe (Plate 5).

PLATE 6
EDIBLE FUNGI IN AFRICA

Photos from the United Republic of Tanzania by Marja Härkönen; Harry Evans for Ghana. All others by Eric Boa.

6.1 (*right*) Roadsides are a common selling point in Malawi. Traders rarely venture beyond markets and collectors must come to them if they chose not to sell themselves.

6.2 (*left*) Made from dried *Uapaca* leaves, this basket is used to store dried mushrooms (and leafy vegetables) collected from the forest. Malawi.

6.3 (*left*) A collector returns with a mornings' harvest. Malawi.

6.4 (*above*) Cleaning a successful harvest (*Termitomyces*). United Republic of Tanzania.

6.5 (*right*) Cooking chanterelles. This mushroom stew is usually eaten with maize or cassava porridge. United Republic of Tanzania.

6.6 (*left*) *Termitomyces* on their way to a local market in Ghana.

6.7 (*above*) Wild edible fungi are also sold dried. United Republic of Tanzania.

6.8 (*right*) Carefully excavating *Termitomyces* in the United Republic of Tanzania. Compare the size with the species from Ghana.

PLATE 7
EDIBLE FUNGI IN LATIN AMERICA AND THE CARIBBEAN

The strong tradition of collecting and eating wild edible fungi extends from Mexico to Guatemala and then appears to stop abruptly. Only one record (shown here) from Bolivia is known. The Caribbean also lacks a tradition of eating wild edible fungi yet, once more, Haitians regularly eat *djon djon* wherever they migrate. Photos from Guatemala by Roberto Flores; New York by Gene Yetter; mushroom fair, Oaxaca by Fabrice Eduard, seller by Elaine Marshall; Bolivia by Eric Boa.

7.1 Mushroom fair to raise awareness of edible species. Oaxaca, Mexico.

7.2 Local market in Oaxaca, Mexico; wild edible fungi shown on right (?*Amanita*) and in front of vendor.

7.3 Patzún market, Guatemala. *Lactarius deliciosus* and *L. indigo* for sale (hand in basket).

7.4 Roadside vendor, Guatemala, with *Lactarius deliciosus* and *Amanita caliptroderma.*

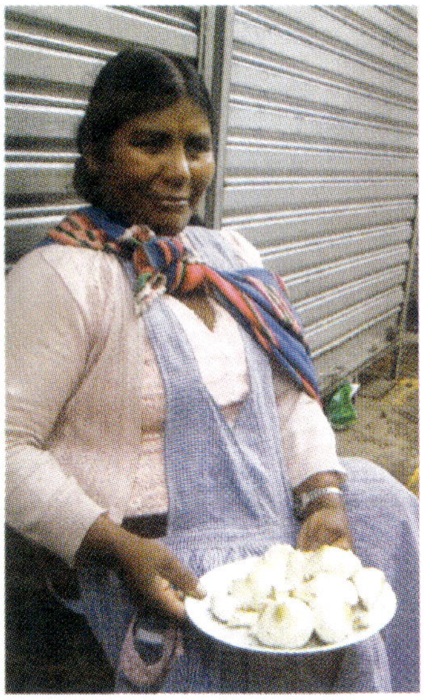

7.5 Gregoria was the only vendor of *k'allampa* (Quechua name for *Leucoagaricus hortensis*) in Cochabamba market, Bolivia.

7.6 Haitian communities around the world regularly buy *djon djon*, a *Psathyrella* sp. Brooklyn, New York.

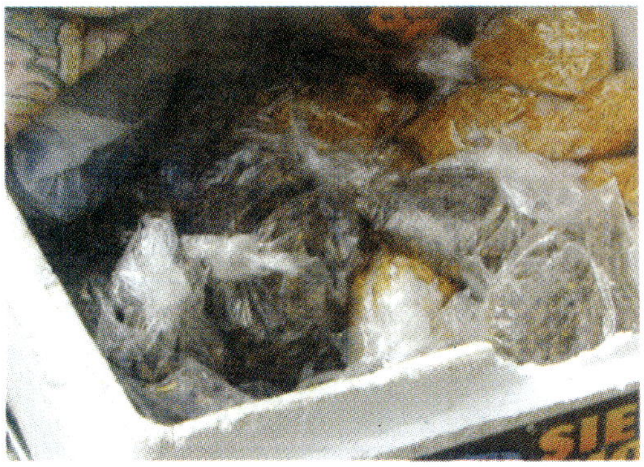

7.7 Fresh *djon djon* are cultivated in Haiti and exported to the United States, Canada and other countries. Brooklyn, New York.

PLATE 8
EDIBLE FUNGI IN ASIA

Photos from Bhutan by Alessandra Zambonelli; Viet Nam and Kunming by Maria Chamberlain; southern China by Marja Härkönen, all other China photos by Warren Priest.

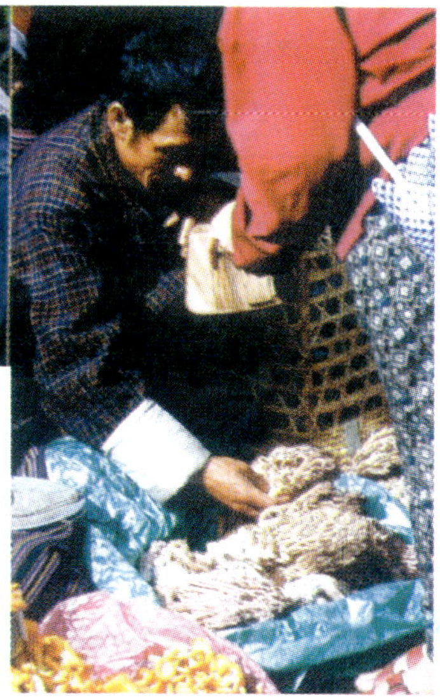

8.1 (*left*) *Matsutake* (*Tricholoma matsutake*) Bhutan.

8.2 (*above*) Preparing *matsutake* for export to Japan from Bhutan. Quality has an important bearing on the prices obtained.

8.3 (*right*) Selling chanterelles and *Ramaria* sp. (in hand), Thimpu, Bhutan

8.4 Cultivating *Agaricus bisporus*, Pohkara, Nepal. Sponsored by Japanese aid project.

8.5 Collector's basket, northern Viet Nam.

8.6 Huge amounts of wild edible fungi are sold in brine. Chengdu, China.

8.7 (*left*) *Termitomyces* are good baby food in Hanyuan county, Sichuan, China.

8.8 (*above*) *Lyophyllum decastes*, for sale in Kunming, China.

8.9 (*right*) A collector in southern China.

5 Realizing the potential: prospects, actions, opportunities

KEY FACTS

The major features of wild edible fungi based on this first global assessment are:

- 2 327 recorded wild useful species; 2 166 are edible and this book has noted 1 069 used as food, with at least 100 other "known food" species still lacking published evidence;
- 470 species have medicinal properties, of which 133 are neither eaten or said to be edible; a further 181 species have other properties and used valued by people, e.g. religious, as tinder;
- they are collected, consumed and sold in over 80 countries worldwide;
- global amount collected each year is several million tonnes with a minimum value of US$2 billion.

The major benefits and features of wild edible fungi, as discussed in Chapters 2 and 4, are:

- they are a valuable source of nutrition, often with associated health benefits;
- they are an important source of income for communities and national economies;
- key species are ectomycorrhizal and help to sustain tree growth and healthy forests;
- they are especially valuable to rural people in developing countries.

GENERAL CONSTRAINTS

Much of the original work on edible fungi has concentrated on the mycological or scientific aspects and, although much still remains to be done, the most significant gaps in information and knowledge concern social and economic aspects of use. Little is known about collectors and collecting practices, for example, or the relative importance of wild edible fungi compared with alternative sources of food or income. Sustainable production of wild edible fungi is not only about how to maximize yields but how to balance this resource with other uses and users of forests.

Despite significant gaps in knowledge it is also important to emphasize that significant advances have been made in describing the features of commercial harvesting in different countries. There is a considerable body of published information from the United States and Canada, for example, and Chinese researchers have also provided new insights concerning the use of wild edible fungi that demonstrate their widespread importance. In central, southern and now west Africa, development projects have explored local use of wild edible fungi while national programmes in Mexico and Turkey have sustained local research programmes over a long period of time.

Now is an appropriate time to identify the most important topics that need further investigation. The following section discusses research priorities in mycology, diet, fungal ecology (mycorrhizas) and storage – how to make better use of annual production. These are key areas where more information is needed. There are many questions about how best to manage wild edible fungi and to achieve sustainable production and this topic is examined in more detail in the subsequent section. Table 25 summarizes the key issues involved and discusses them in relation to commercial harvesting and subsistence uses.

Table 25 and Table 12 attempt to develop a practical approach to management that will be of use to forest managers. The two common constraints for exploring the full potential of wild edible fungi are a poor knowledge of current activities and a lack of reliable data.

RESEARCH PRIORITIES: WILD EDIBLE FUNGI

Identification of species

The tropical mycota is poorly known and concern has been expressed by scientists about the incomplete state of taxonomic knowledge (Meijer, 2001). Steady progress has been made in naming new species of macrofungi (e.g. Verbecken *et al*., 2000; Afyon, 1997) and while there is still much to do there is no obvious evidence that gaps in taxonomic knowledge are limiting the use of wild edible fungi. Local classifications provide a useful guide to edible and "not eaten" species (these may be poisonous or not). Scientific identifications can help to clarify the edibility of species and further information about the identification of macrofungi is always helpful.

The resistance to eating wild fungi is often based on a fear of eating poisonous mushrooms and this does limit the use of edible species and attempts to expand local markets (Lowore and Boa, 2001). Throughout southern Africa *Boletus edulis* is produced in pine plantations but is not eaten locally. Suitable publicity and reassurance from recognized authorities will help to overcome suspicion but concerted efforts are needed to change deep-seated suspicion of wild fungi. Efforts to promote wild edible fungi locally are best concentrated in areas where they are already eaten.

Simple local guides that illustrate useful edible species for a region are more widely needed. Comprehensive field guides are of greatest use but are more costly and complicated to produce. Guides to edible species are not in themselves sufficient: they must be supported by public campaigns that seek to reassure people about which species are safe to eat. The "recognized authorities" refers to both scientists who can identify macrofungi and local people with similar skills acquired from personal experience of what is safe to eat and with a knowledge of local traditions.

Nutritional status

The nutritional benefits of wild edible fungi have not been fully explored. The published information is of variable quality and analytical procedures need to be standardized (Breene, 1990). The range of wild species that have been analysed is still small and little is known about variation within species that occur in different countries, e.g. chanterelles and *Boletus edulis*. Research is needed on species that have greatest market potential and efforts should be made to highlight the nutritional properties and advantages. Many people judge the dietary value of mushrooms with little knowledge of their true properties (see Chapter 2, section on *Edibility and poisonous fungi* and Chapter 4, section on *Nutrition and health benefits* for further information).

Mycorrhizas

The links between wild edible fungi and tree hosts are well known for economically important species such as *Boletus edulis* and *Tuber* spp. *Cantharellus* spp. form mycorrhizae with many tree species in tropical countries. There is an expanding body of information about many other edible fungus–tree associations but this has not been assembled in the form of a database, for example, that would allow for predictive searches. The search for *matsutake* in Asia was assisted by a knowledge of its tree hosts (Namgyel, 2000) and this approach would assist in prospecting for other wild edible fungi. Knowledge about the mycorrhizal partners of edible species of *Amanita*, *Lactarius* and *Russula* is steadily increasing (e.g. Verbecken and Buyck, 2002).

There are potentially large areas of miombo woodland in Malawi which are not accessible to local collectors working on foot, and a better knowledge of which edible mycorrhizal species grow with which trees would help to identify productive areas. In general terms, a database of mycorrhizal associations, linking edible species to tree hosts would help planners and forest managers. The database would need to indicate how well the association had been established. Physical links between macrofungi and trees were relatively simple to trace during one short exercise in Malawi (Plate 2) and published work has already confirmed associations. Even statements such as "found growing in association with" would assist attempts to identify areas where wild edible fungi might occur.

Storage

Wild edible fungi often have a short period during which they can be eaten. They then either rot or shrivel up. They can be preserved in a variety of ways and used at a later date. Some species are readily dried and the flavour of *Boletus edulis* is enhanced by this process (Plate 5). Chanterelles have a longer viable period than many other wild edible species and this enhances their marketability. Truffles also store well, but many other edible fungi are highly perishable. In China, edible fungi are commonly preserved in brine and sold in caskets (Plate 8). They are also exported in this form to Italy.

The technology for preserving wild edible fungi is simple but may require capital investment. Drying mushrooms is more suited to subsistence users and simple methods used in Malawi – dried fungi are stored in natural containers made with dried leaves of *Uapaca kirkiana*, a native tree – have wider applications (Plate 6).

Preserving edible fungi in brine also has wider applications and substantially increases the use and value of wild edible fungi in China. The success of this approach depends on having the equipment and raw materials to carry out the preservation process, but it is important to determine first whether edible fungi in brine are acceptable to the intended market. There is no experience of this method in Africa in rural communities, for example, and market research is needed before contemplating preservation in brine on a wide scale.

Although some wild fungi are dried in southern Africa (Plate 6), there is scope for expanding this approach. If suitable drying methods are not already used, others could be adapted from other areas of agriculture (e.g. drying seeds). It is important in all these efforts to increase the supply of wild edible fungi that they first concentrate on regions where they are already popular and, second, that any new storage methods are developed jointly with local communities.

EFFECTIVE MANAGEMENT

The main objective of managing wild edible fungi is to ensure sustainable production. This is achieved by examining their biology, ecology and patterns of use in relation to other uses of forests and the groups of people involved (Chapter 3). Table 12 outlines the key topics that need to be addressed. Table 25 offers a structured approach towards achieving sustainable production of wild edible fungi and forests.

The key to success is having a sound knowledge of what people do in the forest and why, and assessing the relative importance and priority of benefits obtained (products and services) and related activities. When planning projects or initiatives specifically on wild edible fungi, the objectives of forest management need to be clearly stated: production forests are managed for different purposes compared to protected forests.

The starting point for any management plan is, however, the wild edible fungi themselves. Reliable data are needed on yields and productivity. Recent advice on NWFP inventory methods suggests how this information might be obtained (FAO, 2001a). Lists of species are needed together with information on their relative importance to local people.

Sustainable use of wild edible fungi depends on minimizing the impact of harvesting procedures on the fungus resource and the forest. At the same time, information about

TABLE 25
Information needs and issues concerning sustainable use of wild edible fungi

Key issues	Commercial collections	Personal use/Local sales
Species: which ones are collected	*The range is small and well known. Buyers may require confirmation of species: there are many more tropical species of chanterelles than exist in Europe.* Boletus edulis *from China has a very different flavour to those from Europe.* Hall *et al.*, 2003: general introduction	*The range of species is much greater though not all are of equal importance. Local names can be helpful in overcoming difficulties in naming species. Note the importance of confirming that edible fungi are actually eaten ("food").* De Kesel, Codjia and Yorou, 2002: Benin
Collectors: who are they	*These may be local or from outside. Conflicts occur within and between groups depending on the value of species being collected. The importance of income earned by collectors should be established.* Härkönen, 1998: ethnic groups in Finland	*Mostly for subsistence uses though note collecting for a hobby in the North. Subsistence users vary greatly in social and economic characteristics and this will require careful study.* McLain, Christensen and Shannon, 1998: USA Lowore and Boa, 2001: Malawi
Harvests: how much and impact	*The lure of high prices may lead to the use of harmful methods (both deliberately and unknowingly). Compulsory training exists in the United States and truffle collectors must pass an exam in Italy before being allowed to buy a permit.* Ivancevic, 1997: Yugoslavia	*Harvests are usually small-scale and according to de facto rules established by communities. Data are needed to determine the relative value of collections to rural people. Information on this topic is generally weak.* Malyi, 1987: Belarus
Regulation: use of permits	*Permits are sold in several countries but may prove difficult to monitor. Schemes may need modification and a review of experiences in other countries could be helpful.* Pilz *et al.*, 1999: wild edible fungi, USA	*The concerns are less about the amounts collected than the general presence of collectors in protected forests, leading to concerns about damage to forests and increased risk of fires in some places (USA).* Villarreal and Perez-Moreno, 1989: Mexico
Access: who has rights for collecting	*Commercial harvesting often prompts a closer inspection of who owns or has rights of access to sites. State- or community-run forests are more difficult to manage compared to private plantations.* Yeh, 2000: *matsutake* in China	*The low intensity use associated with personal collections is rarely an issue compared to general concerns about extraction of NWFP from protected forests and conservation areas.* Singh and Rawat, 2000: morels from India
Trade: who buys and sells	*There is a strong imperative for trading systems to develop in a fair and effective manner. Intermediaries are frequently thought to exploit collectors but they also provide credit, a dependable chain for selling and ensure that products get to the market.* Namgyel, 2000: Bhutan	*Markets in southern Africa are small and by the road and this limits the amounts sold. Local trading is often low-key and relatively straightforward.* Lowore and Boa, 2001: Malawi
Yields and productivity: amounts	*The potential threat posed by unsustainable harvests must be determined from an accurate knowledge of yields and productivity data over several years.* Kujala, 1988: Finland	*Yields help to assess the potential for commercialization in local markets.* Vladyshevskiy, Laletin and Vladyshevskiy, 2000: Russian Federation
Markets: amounts traded, exports	*China has a substantial "internal" export market with large amounts flowing from forest to major cities. Elsewhere exports are to Europe and North America. An awareness of relative labour costs determines market opportunities.* www.fintrac.com: export data from several countries	*Market surveys are a useful method for estimating how much is collected locally. They also help to demonstrate the potential for expanding local sales.* Montoya-Esquivel *et al.*, 2001: Mexico; Boa *et al.*, 2000: Malawi
Forest users: who are they and the relative importance of WEF collections	*The collection of high value species may be the main output from a forest and therefore management objectives should be set accordingly.* Tedder, Mitchell and Farran, 2000: Canada	*Rapid appraisal methods have greatly increased knowledge of forest users. Careful analysis of wild edible fungi use is needed – general reports of forest users may not report such practices.* Campbell, 1996: miombo, southern Africa

Forest management: relative importance of wood versus non-wood forest products and specifically wild edible fungi	*A careful examination of forestry objectives with an analysis of major products and services is needed to plan effectively for multiple use.* Alexander *et al.*, 2002: USA	*Low intensity use presents few immediate threats to production forests though a wider knowledge of WEF collecting may alter this current perception.* Lund, Pajari and Korhonen, 1998: boreal and cold temperate forests
Biodiversity: conservation status of wild edible fungi and other plants	*Conservation concerns must address the needs of all forest users, including commercial collections. These cause particular concern because of perceived losses and damage causes. Issues can only be resolved with good and reliable data and a sound understanding of what people do and why.* Perini, 1998: Europe	*A major concern in tropical countries is the poorly described mycota. Studies are currently hampered by a lack of suitably trained taxonomists. A knowledge of ectomycorrhizal associations would help in identifying production of wild edible fungi – as happens with* Tuber *spp. in Europe.* Tibiletti and Zambonelli, 1999: Italy

other forest uses should be gathered. Some uses of a forest may be incompatible and adjustments to their management might be required.

Balancing the needs of forest users in developing countries is often complicated because the pressures on forest resources are great and users have a weak voice in deciding management objectives. User groups must be able to express their needs and feel that their opinions have been taken into account.

COMMERCIALIZATION AND CULTIVATION

Commercialization

There are sometimes unrealistic expectations about money to be earned from exporting wild edible fungi. Much depends on the cost of labour and access to markets. Exports from North America have suffered because harvesting wild edible fungi is cheaper in eastern Europe and transport costs are less. The timing of fruiting seasons will affect the prices that can be achieved. When fruiting seasons overlap in different countries, supplies of common edible species (e.g. chanterelles) will increase and prices will drop. There are yearly fluctuations in production, which are difficult to predict, and fluctuating prices paid for species creates uncertainty and a potentially unstable marketplace.

This is not to say that successful export businesses cannot be sustained, but it requires careful planning, the ability to withstand the ebb and flow of the market place and timely delivery of a good quality product. That is why initiatives to expand local markets are a better way to commercialize wild edible fungi. They will still require attention to detail (getting produce to market quickly) but the potential challenges are smaller and more manageable, thus increasing the chances of success.

Evidence of this comes from local markets in southern Africa and Mexico that have developed out of local initiatives, often with little or no assistance from governments or development projects. The role of researchers and NGOs in these circumstances is to build on existing trading systems and identify where minor changes might lead to major improvements. The following example illustrates the potential of this simple approach.

In Mzimba region in the north of Malawi, women walk long distances in order to meet traders, who buy enthusiastically when the opportunity arises. The strong local demand for wild edible fungi guarantees good market prices yet only a small number of collectors sell their produce directly. More commonly, they sell to the traders who sell in the market at twice the price. Efforts are now being made to encourage more collectors to sell directly and to arrange trading points closer to the collectors' homes, thus increasing the amounts they can supply to local markets (Lowore, Munthali and Boa, 2002).

BOX 9

Product quality and its importance for trade

The roadside sellers of WEF in Malawi are aware that customers will pay more for species that are fresh and presented in an attractive manner. They clean fruiting bodies and select which ones are placed at the tops of piles on their stalls, but on the whole they spend relatively little time in these actions. The differences in money earned are small. The most important thing is to get the WEF as quickly as they can from forest to stall.

As the value of the species increases so too does the increased price that collectors and traders can expect to be paid. The differences in quality between *matsutake* arriving from China and the Republic of Korea in Japan is immediately apparent to anyone comparing boxes. The specimens from the Republic of Korea are less damaged, neatly displayed and in prime condition, thus satisfying the discerning needs of the Japanese customers who will be prepared to pay top prices.

Getting fresh specimens to market is a considerable challenge. The physical appearance of fruiting bodies is obviously important and customer preferences must be observed. Some species discolour if the gills or cap are damaged and they must be handled with care. The buyers have to make sure that fruiting bodies are not infested with insects – some collectors try to hide these at the bottom of trays but such tricks rarely go undetected for long. Depending on the soil where the fungi grow, some preliminary cleaning of gills and gaps may be needed to remove particles. *Sparassis crispa* and other species with honeycomb caps readily accumulate grit, which is difficult to remove.

Picking fruiting bodies at the correct stage of development is important. As they mature some species become woody and much less desirable while others, such as *Coprinus comatus*, quickly dissolve or rot away. The simple consequence for collectors is that inferior specimens are graded lower and are worth less. All things being equal, some provenances of *Boletus edulis* have different taste characteristics. Knowledgeable buyers in Italy can identify the country of origin by smelling the dried fruiting bodies. This in turn determines the price that the buyers will pay for a particular market.

The most spectacular difference in the financial outcomes of product quality is shown by the dramatically different amounts of money earned by the Democratic People's Republic of Korea and the Republic of Korea on exports of *matsutake*. Despite exporting only 264 tonnes over five years, compared to 888 tonnes from the Democratic People's Republic of Korea, the Republic of Korea earned nearly 15 percent more (Tables 22 and 23).

Sources: Lowore and Boa (2001), author's observations and Zambonelli (2002, personal communication: *Truffles, and collecting porcini in Italy*)

Cultivation

There are possibilities for expanding the cultivation of edible fungi. Larger-scale methods are unsuited to local communities that lack the money to establish such businesses. Smaller-scale approaches ("backyard cultivation") are described in Stamets (2000) and widely used throughout China. These have a greater potential for rural people who cultivate paddy-straw as part of integrated farming systems in Viet Nam, for example.

THE FUTURE FOR WILD EDIBLE FUNGI

The increased interest and importance of NWFP have helped to raise the profile of wild edible fungi worldwide. Well-publicized commercial harvesting in North America since the 1990s and the expansion of exports from eastern Europe and China have raised awareness of wild edible fungi and there is now a substantial and significant trade from developing to developed countries. A growing interest in medicinal mushrooms has attracted commercial interests, though there has always been a strong demand in Asia for *Ganoderma* and other key species.

The expansion in commercial harvesting and international trade has led to widespread concern about overharvesting and damage to fungal resources and to forests. There is a danger of restricting commercial harvesting without examining available data or identifying the need to collect data to answer important questions about impact and sustainability. A recent attempt to restrict collections of matsutake in the United States was rejected following a closer look at this resource and its current pattern of use (*Mushroom, the Journal of Wild Mushrooming*, 2002).

The concerns regarding subsistence uses in developing countries are more generally about sustainable use of natural resources. The key to developing wild edible fungi as either a local food or source of income is to examine the different aspects of use and harvesting and to learn more about local practices and community needs.

There has been much enthusiasm for NWFP-based development, particularly in protected forests. Some caution is needed in assessing the potential benefits of this strategy and three commonly held beliefs require closer investigation (Belcher, 2002):

1. NWFP contribute more than timber to the livelihoods and welfare of people living in or near forests, particularly in hard times.
2. Exploitation causes less damage compared with timber harvesting and is a sounder basis for sustainable forest management.
3. Increased commercial harvests add to the value of (tropical) forests and thereby increases the incentive to maintain them rather than convert them to other land uses.

There is better than expected evidence to support the first two points for wild edible fungi while noting the need for more data and better information. It is less clear whether commercial harvests help to protect forests. The mycorrhizal associations of key wild edible fungi do, however, emphasize the unique role they play in maintaining tree health.

The global trade in wild edible (ectomycorrhizal) fungi has been estimated at US$2 billion (Hall *et al.*, 2003). The true value, however, includes the value of wild edible fungi to the millions of rural people around the world who gain benefits from eating them (food they would otherwise have to buy or go without) and money from collecting.

There are compelling reasons for expecting a brighter future for wild edible fungi: they maintain the health of forests; they are a valuable source of nutrition and income. New initiatives should concentrate on expanded use and benefits in areas that already have a strong tradition of wild edible fungi. Export opportunities also exist but are inherently more risky.

During the preparation of this book information on wild edible and wild useful fungi was stored in a simple database. This has been extensively updated and modified with the assistance of Dr Paul Kirk of CABI Bioscience and can be queried over the Internet (www.wildusefulfungi.org). Summary information on over 2 600 species is available and the original records from over 1 000 references and lists published around the world can be viewed. This new Web site also provides a simple means for checking valid and preferred names of WEF species.

PLATE 9
EDIBLE AND MEDICINAL FUNGI IN ASIA

All photos by Eric Boa except *Cordyceps sinensis* photos by Warren Priest.

9.1 Packaging for *Phallus impudicus.*

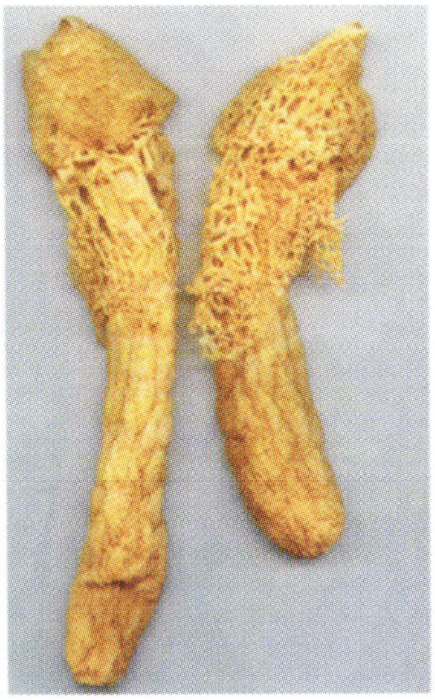

9.2 Dried *Phallus impudicus.*

9.3 Dried morels, bought in Belgium.

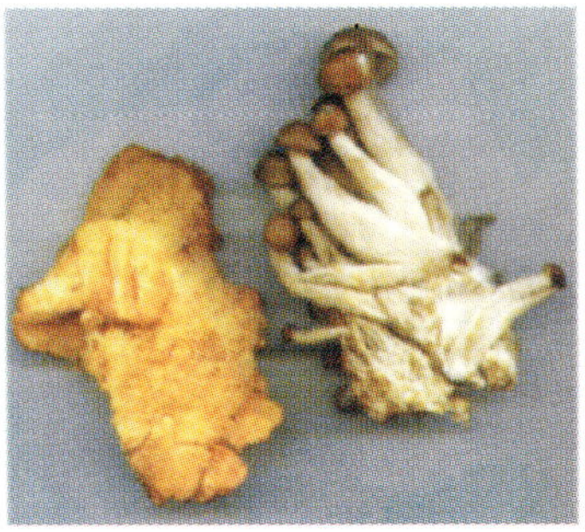

9.4 (*left*) Dried *Cantharellus cibarius* for sale in Hungary.

9.5 (*above*) Fresh *Hydnum repandum* (left – note spines, *sans* gills) and *Hypsizygus tessulatus* for sale in a UK supermarket.

9.6 (*right*) *Ganoderma*, dried, sold for medicinal purposes. Singapore.

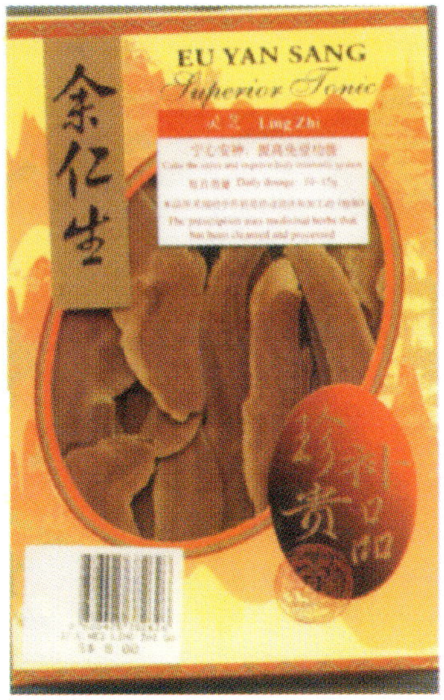

9.7 Shops advertise *chongcao* (*Cordyceps sinensis*) – the orange "sticks" on the left – in Xining, China.

9.8 Cleaning *chongcao* in Kangding, China in preparation for selling.

6 Sources of advice and information

MYCOLOGICAL EXPERTISE

One of the most common areas where technical advice is sought is in identifying specimens and obtaining a scientific name. There are mycologists in all major countries, both developed and developing, though their experience of macrofungi may be limited to particular groups. Many mycologists work with microfungi and in other applied areas such as plant pathology.

Experts on edible fungi are likely to be most knowledgeable about the cultivated species. Wild edible fungi have not been the focus of concerted research until the last ten or twenty years and professional expertise is subject to the vagaries of short-term funding, particularly when it comes to the study of subsistence uses. Individual researchers maintain a close professional interest in wild edible fungi, though this is often broad-based and not specialized in the identification of species.

There are, however, various professional groups with a shared interest in edible fungi which meet on a regular basis. Individual members are dispersed around the world. The best known example is the Edible Ectomycorrhizal Group, which can be contacted via a Web site listed in Table 28.

There are a number of institutes based in Europe and North America which have an international outreach and these are listed below. The major herbaria where reference collections of macrofungi are stored are based in developed countries, although efforts are being made to establish collections elsewhere. Mycological expertise in identifying specimens is available in major countries such as Mexico and China. It is not always clear which institute or individual might be able to assist with identifications and the best general advice is to look via general Web sites or Internet search engines.

On the wider issues of NWFP, ethnoscience, participatory approaches to development and other disciplines relevant to the use of wild edible fungi, FAO is a good starting point for assistance.

Mycological societies exist in many different countries and are a useful starting point for enquiries (see Table 28 for details of Web sites).

FIELD GUIDES TO WILD (EDIBLE) FUNGI

There are many field guides to macrofungi, which include information on edible and poisonous species. They are intended for naturalists and people who go collecting for the occasional mushroom to eat. Detailed field guides contain scientific descriptions of species, expressed in a concise and unambiguous language that is often difficult for the non-specialist to understand. Shorter pocketbooks are available which rely more on photographs and have only short written descriptions of species. Both types of guide are useful for identifying species but they are mostly written for audiences in developed countries and have, therefore, a limited use in developing countries.

There are few books that address the topic of wild edible fungi specifically from a people perspective and most of the relevant information is scattered across a wide range of disciplines (see Table 2 for more information). The best general introduction on wild edible fungi, including helpful details about uses, is a book first published in New Zealand (Hall *et al.*, 1998a). A new edition was published in 2003 (Hall *et al.*,

TABLE 26
Sources of technical advice and information on wild edible fungi

Organization	Contact details	Notes
CABI *Bioscience*	Bakeham Lane Egham Surrey TW20 9TY United Kingdom	Incorporates the International Mycological Institute; herbarium; publications; reference library; taxonomic expertise; broad development experience; databases and Index Fungorum. www.wildusefulfungi.org; www.cabi-bioscience.org
Royal Botanic Gardens, Kew	The Herbarium Surrey TW9 3AB United Kingdom	Herbarium; taxonomic expertise in macrofungi; centre for Economic Botany (including edible fungi); reference library. www.rbgkew.org.uk/scihort/mycolexp.htm
National Museum Belgium	Domein van Bouchot B-1860 Meise Belgium	Taxonomic expertise; wild edible fungi; herbarium, international links; publications. www.br.fgov.be
Crop and Food Research Institute	PB 470 Christchurch New Zealand	Technology development. Growing truffles and other wild edible fungi in "managed" conditions. www.crop.cri.nz/psp/em-mushrooms/index.htm

TABLE 27
Field guides and Web sites for identifying macrofungi and edible varieties

Country	Information and source
Argentina	*Gamundí and Horak, 1995*: macrofungi, pocketbook with colour photos. In Spanish.
Benin	*De Kesel, Codjia and Yorou., 2002:* selected photographs, species descriptions. In French.
Bulgaria	*Iordanov, Vanev and Fakirova, 1978:* edible and poisonous species, in Bulgarian. Drawings.
Burundi	*Buyck, 1994b:* annotated guide to edible species In French. Photographs.
China	The most cinorehensive and best illustrated guide is *Mao, 2000,* a stunning compendium of field mycology with extensive colour photographs. *Ying* et al., *1988*: edible species, in Chinese [not seen]. *Mao, 1998*: Edible species, in Chinese. *Ying* et al., *1987*: medicinal species, in Chinese [not seen]. www.im.ac.cn: has photographs of major economic species.
Colombia	*Franco-Molano, Aldana-Gomez and Halling, 2000*: guide to macrofungi, photographs.
Costa Rica	Two excellent guides with good colour photographs and Spanish and English text are available (*Mata*, 2003; *Halling and Mueller*, 2003).
India	*Purkayastha and Chandra, 1985:* useful summary of edible species, nutrition data. No photographs or drawings.
Israel	*Wasser, 1995:* edible and poisonous species, in Russian and Hebrew [not seen].
Italy	*Testi, 1999* is a popular guide, one of many published. Edible fungi from Basilicate are described in *Tagliavini and Tagliavini, 2001.* Both guides have photographs and are in Italian.
Japan	*Imazeki* et al., *1988*: fungi of Japan, in Japanese but species names in English and many fine photos.
Korea (Republic of)	*Park and Lee, 1999:* guide to Korean mushrooms. Not seen – in Korean.
Kyrgyzstan	*El'chibaev, 1964:* edible mushrooms, drawings, in Russian.
Lao People's Democratic Republic	*http://giechgroup.hp.infoseek.co.jp/kinoko/eng.html*: mostly photographs, limited text.
Malawi	*www.malawifungi.org*: edible species, with photographs, reports and database of local names. *Morris, 1987:* edible species. Drawings.
Mexico	*www.semarnat.gob.mx*: edible, poisonous and medicinal species, in Spanish. Text and photographs.
Poland	*www. grzyby.pl*: brief guide to commercial species, with photographs, in Polish and English.
Russian Federation (far east)	*Vasil'eva, 1978:* edible, poisonous and medicinal species, in Russian, seen only in translation. There are many popular guides to field mushrooms, and the following is a useful and readily available example. It is in Russian and has drawings: *Sergeeva, 2000.* .
Southern Africa	*Ryvarden, Piearce and Masuka., 1994:* describes macrofungi in general, including edible species. Photographs. *van -der -Westhuizen and Eicker, 1994:* general guide to macrofungi, photographs and species descriptions of most relevance to South Africa.
Spain	*Rodriguez* et al. *(1999)* macrofungi with notes on edibility, colour photos, in Spanish.
Tanzania (United Republic of)	*Härkönen, Niemelä and Mwasumbi,* 2003.
Tibet Autonomous Region,China	*Mao and Jiang, 1992*: Economic macrofungi, in Chinese [not seen].
Turkey	*www.ogm.gov.tr/:* edible species, in English. Photographs and short text.
Uganda	*Katende, Segawa and Birnie, 1999:* limited range of edible species, drawings.
Ukraine	*Zerova and Rozhenko, 1988:* edible and poisonous species, in Russian. Drawings. *Wasser, 1990:* guide to edible and poisonous species of Carpathians.
United Kingdom	*Phillips* et al., *1983*: edible and poisonous species, excellent photographs.
United States	*Arora, 1986:* popular guide to all macrofungi with many photographs. *www.mykoweb.com:* edible species, photographs, descriptions. *Molina* et al., *1993:* major edible species in Pacific northwest, photographs.

TABLE 28
General Web sites on wild edible fungi and related topics

ADDRESS	COMMENTS
http://mycology.cornell.edu	Virtual Library on Mycology. Main po[illegible] information on fungi, including useful species. Good starting point for general enquiries.
www.mushworld.com	One of the most useful of many "commercial" sites investigated. Access is free once you have registered. Has reports on mushroom production (cultivated) and has a good global coverage.
http://.mycorrhiza.ag.utk.edu	International Directory of Mycorrhizologists. Links to sites on edible ectomycorrhizal mushrooms, lists scientists and has many other useful background information. Good general reference point.
www.indexfungorum.org	Essential reference tool. Check species names of all fungi, including macrofungi, and also the correct authorities.
http://gmr.landfood.unimelb.edu.au/~plants/	Multilingual guide to fungus names, including Chinese. Does not have a special emphasis on wild fungi.
www.malawifungi.org	Wild useful fungi of Malawi with a searchable database of local names and scientific equivalents. Project reports can be downloaded; photographs of many species are available.
www.im.ac.cn	Economic fungi of China. Many photographs; wayward spellings of scientific names.
www.semarnat.gob.mx	Excellent site (in Spanish) giving details of major wild edible fungi from Mexico, including full descriptions and photos.
www.grzyby.pl	Edible fungi of Poland (some text in English).
http://fungimap.rgb.vic.gov.au	General information on edible and poisonous species of Australia.
www.fintrac.com	Contains useful trade data from 1993– 97 for "mushroom" exports to selected countries and specifically for matsutake exports to Japan.
www.fungi.com	Fungi Perfecti, a commercial company specializing in the cultivation of gourmet and medicinal mushrooms. Good general information and many links.
www.mycopat.slu.se/mycorrhiza/edible/home.phtml	Edible mycorrhizal mushrooms. Two international conferences have been held and the site gives information on talks and other matters of general relevance to WEF.
www.mushroomthejournal.com	The journal of wild mushrooming, published in the United States with articles available online. Presents a very practical approach and analysis of mushroom collecting and although slanted towards the amateur in the United States, it explores universal issues (regulation of collectors) of broader relevance.
www.fs.fed.us	Information on commercial harvesting in the Pacific northwest of the United States, including detailed accounts from Winema National Forest.
www.wildusefulfungi.org	

2003). A dictionary of edible fungi contains lists of species from several developed and developing countries and local names. It is a useful but not essential reference (Chandra, 1989).

Country guides

Most field guides are based on species found in temperate regions. There is a plethora of such guides from the United States while countries in western Europe are also well served. Key examples are listed in Table 27 but the emphasis is on less well known books from developing countries. Most are out of print and only available from specialist libraries. Guides published in the United States (e.g. Arora, 1986) and Europe (e.g. Phillips *et al.*, 1983) can still be purchased or readily consulted in libraries.

INFORMATION ON MEDICINAL AND POISONOUS MUSHROOMS

Many edible fungi also have medicinal properties. The *International Journal of Medicinal Mushrooms* began publication in 1999 and contains review articles as well as original contributions. For a general overview see Hobbs (1995).

All guides to macrofungi include descriptions of poisonous species. There is a colour atlas devoted to poisonous species though the examples are of species found in developed countries, some of which will also occur in developing countries (Bresinsky and Besl, 1990).

WEB SITES

The Internet is a useful source of information but the quality and accuracy of this information can be difficult to assess. Type the word "mushroom" or "edible fungus" into a search engine such as Google (www.google.com) and a barrage of Web addresses will appear. The sites listed in Table 28 are a starting point for investigations and notes have been provided to indicate how useful they were during the preparation of this book. Most sites listed in Table 28 emphasize fungi first and uses by people second – if at all.

Table 28 is only a selection of available Web sites that include wild edible fungi. For more detailed searches of reliably published information there is no substitute for thorough literature reviews of journals and other professionally published sources. Table 28 includes examples of country-specific Web sites, and attention is drawn to the excellent information available for Mexico.

7 References

Aaronson, S. 2000. Fungi. *In* K.F. Kiple & K.C. Ornelas, eds. *The Cambridge world history of food*, pp 313–336. Cambridge, UK, Cambridge University Press. 1 958 pp.

Abate, D. 1999. *Agaricus campestris* in upland Ethiopia. *Mycologist,* 13: 28.

Abbott, P. 1999. Non-timber forest products harvesting: lessons for seasonally-sensitive management in miombo. *In* M.R. Ngulube, L. Mwabumba & P. Chirwa, eds. *Community-based management of miombo woodlands in Malawi*, pp. 70–89. Proceedings of a national workshop, Sun and Sand Holiday Resort, Mangochi, Malawi, 27 to 29 September 1999. Zomba, Malawi, Forestry Research Institute of Malawi.

Adhikari, K.S. & Adhikari, M.K. 1996. Collection and consumption of wild edible mushrooms sold in Kathmandu valley, Nepal. *The Geographer's Point,* 1–2: 1–9.

Adhikari, M.K. 1999. Wild relatives of some arable mushrooms found in Nepal. *In* National Conference on Wild Relatives of Cultivated Plants in Nepal, pp. 149–155. Kathmandu, Green Energy Mission.

Adhikari, M.K. & Durrieu, G. 1996. Ethnomycologie Nepalaise. *Bulletin Societé Mycologique de France,* 112: 31–41.

Adhikary, R.K., Baruah, P., Kalita, P. & Bordoloi, D. 1999. Edible mushrooms growing in the forests of Arunachal Pradesh. *Advances in Horticulture and Forestry,* 6: 119–123.

Afyon, A. 1997. Macrofungi of Seydisehir district (Konya). *Turkish Journal of Botany,* 21(3): 173–176.

Ainsworth, G.C. 1976. *Introduction to the history of mycology*. Cambridge, UK, Cambridge University Press. 359 pp.

Alexander, I.J. & Hogberg, P. 1986. Ectomycorrhizas of tropical angiospermous trees. *New Phytologist,* 102: 541–549.

Alexander, S., Pilz, D., Weber, N.S., Brown, E. & Rockwell, V.A. 2002. Mushrooms, trees and money: value estimates of commercial mushrooms and timber in the Pacific Northwest. *Environmental Management,* 30: 129–141.

Alexiades, M.N., ed. 1996. *Selected guidelines for ethnobotanical research: a field manual.* New York, USA, New York Botanical Garden. 306 pp.

Almond, M. 2002. Eddie George takes over Ukraine. *New Statesman,* 8 April 2002: 31.

Al-Naama, N.M., Ewaze, J.O. & Nema, J.H. 1988. Chemical constituents of Iraqi truffles. *Iraqi Journal of Agricultural Sciences,* 6: 51–56.

Alofe, F.V., Odeyemi, O. & Oke, O.L. 1996. Three edible wild mushrooms from Nigeria: their proximate and mineral composition. *Plant Foods for Human Nutrition,* 49: 63–73.

Alphonse, M.E. 1981. *Les champignons comestible d'Haiti.* Port au Prince. (publisher not known)

Alsheikh, A.M. & Trappe, J.M. 1983. Desert truffles: the genus *Tirmania. Transactions of the British Mycological Society,* 81: 83–90.

Antonin, V. & Fraiture, A. 1998. *Marasmius heinemannianus*, a new edible species from Benin, West Africa. *Belgian Journal of Botany,* 131: 127–132.

Arnolds, E. 1995. Conservation and management of natural populations of edible fungi. *Canadian Journal of Botany,* 73: 987–998.

Aroche, R.M., Cifuentes, J., Lorea, F., Fuentes, P., Bonavides, J., Galicia, H., Menendez, E., Aguilar, O. & Valenzuela, V. 1984. Poisonous and edible macromycetes in a communal region of the Valle de Mexico, I. *Boletin de la Sociedad Mexicana de Micología,* 19: 291–318.

Arora, D. 1986. *Mushrooms demystified.* Berkeley, CA, USA, Ten Speed Press. 420 pp.

Arora, D. 1999. The way of the wild mushroom. *California Wild,* 52(4): 8–19.

Bandala, V.M., Montoya, L. & Chapela, I.H. 1997. Wild edible mushrooms in Mexico: a challenge and opportunity for sustainable development. *In* M.E. Palm & I.H. Chapela, eds. *Mycology in sustainable development: expanding concepts, vanishing borders,* pp. 76–90. Boone, NC, USA, Parkway Publishers.

Baptista-Ferreira, J.L. 1997. What's going on about conservation of fungi in Portugal. *In* C. Perini, ed. *Conservation of fungi,* pp. 35–37. Siena, Italy, Universita degli Studi di Siena.

Barnett, C.L., Beresford, N.A., Frankland, J.C., Self, P.L., Howard, B.J. & Marriott, J.V.R. 2001. Radiocaesium intake in Great Britain as a consequence of the consumption of wild fungi. *Mycologist,* 15(3): 98–104.

Batra, L.R. 1983. Edible discomycetes and gasteromycetes of Afghanistan, Pakistan and north-western India. *Biologia (Lahore),* 29: 293–304.

Bauer-Petrovska, B., Jordanoski, B., Stefov, V. & Kulevanova, S. 2001. Investigation of dietary fibre in some edible mushrooms from Macedonia. *Nutrition and Food Science,* 31: 242–246.

Belcher, B. 2002. CIFOR research: forest products and people-rattan issues. *In* FAO Non-wood Forest Products 14. *Rattan. Current research issues and prospects for conservation and sustainable development*, pp. 49–62. J. Dransfield, F.O. Tesoro & N. Manokaran, eds. Rome, FAO.

Bérelle, G. 2002. Organiser le ramassage des champignons (Organizing the collection of mushrooms). *Forets de France,* 456: 31.

Birks, A.A. 1991. Fungi in folk medicine. *McIlvainea,* 10(1): 89–94.

Boa, E.R. 2002. How do local people make use of wild edible fungi? Personal narratives from Malawi. *In* I.R. Hall, Y. Wang, A. Zambonelli & E. Danell, eds. *Edible ectomycorrhizal mushrooms and their cultivation.* Proceedings of the second international conference on edible ectomycorrhizal mushrooms. July 2001, Christchurch. CD-ROM. Christchurch, New Zealand Institute for Crop and Food Research Limited.

Boa, E.R., Ngulube, M. Meke, G. & Munthali, C. eds. 2000. *First Regional Workshop on Sustainable Use of Forest Products: Miombo Wild Edible Fungi.* Zomba, Malawi, Forest Research Institute of Malawi and CABI Bioscience. 61 pp.

Bokhary, H.A. & Parvez, S. 1993. Chemical composition of desert truffles *Terfezia claveryi. Journal of Food Composition and Analysis,* 6(3): 285–293.

Boruah, P., Adhikary, R.K., Kalita, P. & Bordoloi, D. 1996. Some edible fungi growing in the forest of East Khasi Hills (Meghalaya). *Advances in Forestry Research in India,* 14: 214–219.

Boruah, P. & Singh, R.S. 2001. Edible fungi of medicinal value from the eastern Himalaya region. *International Journal of Medicinal Mushrooms,* 3: 124.

Bouriquet, G. 1970. Les principaux champignons de Madagascar. *Terre Malagache,* 7: 10–37.

Breene, W.M. 1990. Nutritional and medicinal value of specialty mushrooms. *Journal of Food Protection,* 53: 883–894.

Bresinsky, A. & Besl, H. 1990. *A colour atlas of poisonous fungi.* London, Wolfe.

Bukowski, T. 1960. Mushroom growing in Poland. In *Mushroom Science IV.* pp. 504–506. Proceedings of the fourth international conference on scientific aspects of mushroom growing, 18–26 July 1959, Copenhagen. Odense, Denmark, Andelsbogtrykkeriet.

Bulakh, E.M. 2001. Medicinal mushrooms of the Russian far east in nature. *International Journal of Medicinal Mushrooms,* 3: 125.

Buller, A.H.R. 1914. The fungus lores of the Greeks and Romans. *Transactions of the British Mycological Society,* 5: 21–66.

Burkhill, I.H. 1935. *A dictionary of the economic products of the Malay Peninsula.* London, Crown Agents for the Colonies.

Butkus, V., Jaskonis, I. Urbonas, V. & Cervokas, V. 1987. *Minor forest resources: fruit bearing plants; medicinal plants; fungi.* Lithuania. 415 pp.

Buyck, B. 1994a. Ectotrophy in tropical African ecosystems. *In* J.H. Seyani & A.C.

Chikuni, eds. *Proceedings of the XIIIth Plenary meeting of AETFAT*, pp. 705–718. Zomba, Malawi, 2–11 April 1991. Zomba, Malawi, National Herbarium and Botanic Gardens of Malawi.

Buyck, B. 1994b. *Ubwoba: Les champignons comestibles de l'ouest du Burundi*. Brussels, Administration Generale de la Cooperation au Developpement. 123 pp.

Buyck, B. 2001. Preliminary observations on the diversity and habitats of Russula (Russulales, Basidiomycotina) in Madagascar. *Micologia e Vegetazione Mediterranea*, 16: 133–147.

Caglarirmak, N., Unal, K. & Otles, S. 2002. Nutritional value of wild edible mushrooms collected from the Black Sea region of Turkey. *Micologia Aplicada International*, 14(1): 1–5.

Campbell, B., ed. 1996. *The miombo in transition: woodlands and welfare in Africa*. Bogor, Indonesia, Centre for International Forestry Research. 266 pp.

Cao, J. 1991. A new wild edible fungus - *Wynnella silvicola. Zhongguo Shiyongjun (Edible fungi of China: a bimonthly journal)*, 10(1): 27–28.

Cavalcaselle, B. 1997. Edible mushroom production in forest villages of Turkey, Syria and Jordan. In *Medicinal and culinary plants in the Near East*, Proceedings of the international expert meeting. Cairo, FAO.

Cervera, M. &Colinas, C. 1997. Comercializazión de seta silvestre en la ciudad de Lleida. *In* F. Puertas & M. Rivas, eds. *Actas del I Congreso Forestal Hispano LUso, II Congreso Forestal Espanol-IRATI 97*, pp. 425–429. Pamplona, Spain, 23–27 June 1997.

Chamberlain, M. 1996. Ethnomycological experiences in South West China. *Mycologist*, 10: 13–16.

Chandra, A. 1989. *Elsevier's dictionary of edible mushrooms. Botanical and common names in various languages of the world*. Amsterdam, Netherlands, Elsevier. 259 pp.

Chang, S.T. 1991. Mushroom biology and mushroom production. *Mushroom Journal for the Tropics*, 11: 45–52.

Chang, S.T. 1999. World production of cultivated edible and medicinal mushrooms in 1997 with emphasis on *Lentinus edodes* in China. *International Journal of Medicinal Mushrooms*, 1: 291–300.

Chang, S.T. & Buswell, J.A. 1999. *Ganoderma lucidum* – a mushrooming medicinal mushroom. *International Journal of Medicinal Mushrooms*, 1: 139–146.

Chang, S.T. & Mao, X. 1995. *Hong Kong mushrooms*. Hong Kong, Chinese University of Hong Kong. 470 pp.

Chang, S.T. & Miles, P.G. 1991. Recent trends in world production of cultivated edible mushrooms. *Mushroom Journal*, 504: 15–18.

Chen, Z.C. 1987. Distribution of Agaricales in Taiwan. *Transactions of the Mycological Society of Republic of China*, 2(1): 1–21.

Chibisov, G. & Demidova, N. 1998. Non-wood forest products and their research in Arkhangelsk, Russia. *In* H.G. Lund, B. Pajari & M. Korhonen, eds. *Sustainable development of non-wood goods and benefits from boreal and cold temperate forests*, pp. 147–153. Proceedings of the International Workshop, Joensuu, Finland, 18–22 January 1998. EFI-Proceedings. 1998, No. 23.

Chin, F.H. 1988a. Edible and poisonous fungi from the forests of Sarawak. Part 1. *The Sarawak Museum Journal*, 29: 211–225.

Chin, F.H. 1998b. Edible and poisonous fungi from the forests of Sarawak. Part 2. *The Sarawak Museum Journal*, 39: 195–201.

Cochran, K.W. 1987. Poisonings due to misidentified mushrooms. *McIlvainea*, 8(1): 27–29.

Colenso, W. 1884–85. On a New Zealand fungus that has of late years become a valuable article of commerce. *Report and Transactions of the Penzance Natural History and Antiquarian Society*, pp 82–86.

Crisan, E.V. & Sands, A. 1978. Nutritional value. *In* S.T. Chang & W.A.Hayes, eds. *The biology and cultivation of edible fungi*, pp. 137–168. New York, USA, Academic Press.

Davis, J. 2000. The edible and medicinal mushrooms industry in Australia. *International Journal of Medicinal Mushrooms,* 2(1): 5–9.

Davis, W. 1996. *One River*. New York, USA, Simon and Schuster. 537 pp.

de Beer, J. & Zakharenkov, A. 1999. Tigers, mushrooms and bonanzas in the Russian far east: the Udege 's campaign for economic survival and conservation. *In* P. Wolvekamp, A.D. Usher, V. Paranjpye & M. Ramnath, eds. *Forests for the future: local strategies for forest protection, economic welfare and social justice,* pp. 244–250. London, Zed Books.

de Geus, N. 1995. *Botanical forest products in British Columbia. An overview.* Victoria, BC, Canada, British Columbia Ministry of Forests.

Degreef, J. 1992. *Inventaire, valeur alimentaire et culture de champignons du Shaba.* Memoire fin d'etudes, Chaire d'Ecologie et Phytosociologie, Fac. Sciences Agron. de Gembloux, Belgium.

Degreef, J., Malaisse, F., Rammeloo, J. & Baudart, E. 1997. Edible mushrooms of the Zambezian woodland area: a nutritional and ecological approach. *BASE (Biotechnologie, Agronomie, Societe et Environnement),* 1: 221–231.

De Kesel, A., Codjia, J.T.C. & Yorou, S.N. 2002. *Guide des champignons comestibles du Bénin.* Cotonou, République du Bénin, Jardin Botanique National de Belgique et Centre International d'Ecodéveloppement Intégré (CECODI. Impr. Coco-Multimedia. 275 pp.

de Leon, R. 2002. Cultivated edible and medicinal mushrooms in Guatemala (available at www.mushworld.com).

Demirbas, A. 2000. Accumulation of heavy metals in some edible mushrooms from Turkey. *Food Chemistry,* 68: 415–419.

Deschamps, J.R. 2002. *Hongos silvestres comestibles del Mercosur con valor gastronómico,* Documentos de trabajo. No. 86. Universidad de Belgrano, Argentina. 25 pp.

Diamandis, S. 1997. Conservation of fungi in Greece. *In* C. Perini, ed. *Conservation of Fungi,* pp. 44–46. Siena, Italy, Universita degli Studi di Siena.

Didukh, I.A. 2001. Mushrooms in folk medicine of the eastern Slavs. *International Journal of Medicinal Mushrooms,* 3: 135.

Dong, M. & Shen, A. 1993. Studies on *Lactarius camphoratus.* 1 Biological characteristics of *L. camphoratus. Zhongguo Shiyongjun (Edible fungi of China: a bimonthly journal),* 12(1): 3–5.

Ducousso, M., Ba, A.M. & Thoen, D. 2002. Ectomycorrhizal fungi associated with native and planted tree species in West Africa: a potential source of edible mushrooms. *In* I.R. Hall, Y. Wang, A. Zambonelli & E. Danell, eds. *Edible ectomycorrhizal mushrooms and their cultivation.* Proceedings of the second international conference on edible mycorrhizal mushrooms. July 2001, Christchurch. CD-ROM. Christchurch, New Zealand Institute for Crop and Food Research Limited.

Dudka, I.A. & Wasser, S.P. 1987. *Mushrooms. A reference book for the mycologist and the mushroom collector (in Russian).* Kiev, Naukova Dumka Press. 536 pp.

Dyke, A.J. & Newton, A.C. 1999. Commercial harvesting of wild mushrooms in Scottish forests: is it sustainable? *Scottish Forestry,* 53: 77–85.

Egli, S., Ayer, F. & Chatelain, F. 1990. Die Beschreibung der Diversitat von Makromyzeten. Erfahrungen aus pilzokologischen Langzeitstudien im Pilzreservat La Chanéaz, FR. *Mycologia Helvetica,* 9(2): 19–32.

El'chibaev, A.A. 1964. *S'edobnye griby Kirgizii [Edible mushrooms of the Kirghiz SSR],* Kirgizskoi SSR, Izdatel'stvo Akademii Nauk. 44 pp.

Ereifej, K.I. & Al-Raddad, A.M. 2000. Identification and quality evaluation of two wild mushrooms in relation to *Agaricus bisporus* from Jordan. *In* L.Van Griensven, ed. *Science and cultivation of edible fungi,* pp. 721–724. Proceedings of the 15th International Congress on the Science and Cultivation of Edible Fungi, Maastricht, Netherlands, 15–19 May 2000.

Ertrug, F. 2000. An ethnobotanical study in Central Anatolia (Turkey). *Economic Botany,* 54(2): 155–182.

Etkin, N.L. & Johns, T. 1998. 'Pharmafoods' and 'Nutriceuticals': paradigm shifts in biotherapeutics. *In* H.V. Prendergast, N.L. Etkin, D.R. Harris & P.J. Houghton, eds. *Plants for food and medicine*, pp. 3–16. Kew, London, Royal Botanic Gardens.

Evans, L. 1996. Why so many poisonings in Russia? (letter). *Journal of Wild Mushrooming* **14**(1): 4.

FAO. 1993a. *Cosecha de hongos en la VII región de Chile.* Estudio monografico de explotación forestal – 2. Rome, FAO. 35 pp.

FAO. 1993b. *International trade in non-wood forest products: an overview,* by M. Iqbal. Working Paper Misc/93/11. Rome, 100 pp.

FAO. 1998a. *Principales productos forestales no madereros en Chile,* by J. Campos. Santiago, Chile.

FAO. 1998b. *Non-wood forest products from conifers,* by W.M. Ciesla. Non-wood Forest Products 12. Rome. 138 pp.

FAO. 2001a. *Resource assessment of non-wood forest products: experience and biometric principles.* by J. Wong, K. Thornber & N. Baker. Non-wood Forest Products 13. Rome. 126 pp.

FAO. 2001b. *Non-wood forest products in Africa: a regional and national overview,* by S. Walter. Rome. 303 pp.

Federation-Francaisc-des-Trufficulteurs. 2001. *Science et culture de la truffe.* Actes du Ve Congres International, 4–6 March 1999, Aix-en-Provence, France. 563 pp.

Fidalgo, O. & Prance, G.T. 1976. The ethnomycology of the Sanama Indians. *Mycologia,* 68: 201–210.

Filipov, D. 1998. Mushroom season has Russians in fungi frenzy. *Boston Globe,* 6 September, 1998.

Flores, R., Bran, M.d.C. & Honrubia, M. 2002. Edible mycorrhizal mushrooms of the west Highland Guatemala. *In* I.R. Hall, Y. Wang, A. Zambonelli & E. Danell, eds. *Edible ectomycorrhizal mushrooms and their cultivation.* Proceedings of the second international conference on edible mycorrhizal mushrooms. July 2001, Christchurch. CD-ROM. Christchurch, New Zealand Institute for Crop and Food Research Limited.

Franco-Molano, A., Aldana-Gomez, R. & Halling, R.E. 2000. *Setas de Colombia (Agaricales, Boletales y otras Hongos. Guía de Campo.* Medellín, Colombia, COLCIENCIAS, Universidad de Antioquia. 156 pp.

Franquemont, C., Plowman, T., Franquemont, E., King, S.R., Niezgoda, C., Davis, W. & Sperling, C.R. 1990. The ethnobotany of Chinchero, an Andean community in southern Peru. *Fieldiana,* 24: 1–126.

Gamundí, I. & Horak, E. 1995. *Fungi of the Andean-Patagonian forests.* Buenos Aires, Vazquez Mazzini Editores. 141 pp.

Gardezi, R.A. 1993. Agaric fungi from Rawalakot, Azad Kashmir. *Sarhad Journal of Agriculture,* 8(3): 225–226.

Gecan, J.S. & Cichowicz, S.M. 1993. Toxic mushroom contamination of wild mushrooms in commercial distribution. *Journal of Food Protection,* 56(8): 730–734.

Gong, C.L. & Peng, G.P. 1993. Culture of *Cordyceps militaris* on Chinese silkworms and the analysis of its components. *Zhongguo Shiyongjun (Edible Fungi of China: a bimonthly journal),* 12(4): 21–23.

Grunert, H. & Grunert, R. 1995. *Gombák [Mushrooms].* Budapest, Magyar Konyvklub.

Grzymala, S. 1965. Les recherches sur la frequence des intoxications par les champignons. *Bull. Med. Legale,* 8(2): 200–210.

Grzywnowicz, K. 2001. Medicinal mushrooms in Polish folk medicine. *International Journal of Medicinal Mushrooms,* 3: 154.

Gunatilleke, I.A.U.N., Gunatilleke, C.V.S. & Abeygunawardena, P. 1993. Interdisciplinary research towards management of non-timber forest resources in lowland rain forests of Sri Lanka. *Economic Botany,* 47(3): 282–290.

Gunawan, A.W. 2000. *Usaha pembibitan jamur [growing mushrooms].* Jakarta, Penebar Swadaya. 112 pp.

Guzmán, G. 1997. *Los nombres de los hongos y lo relacionado con ellos en América Latina:*

introducción a la entomicabiota y micología aplicada de la región (Sinonimia vulgar y científica. Jalapa, Veracruz, CONABIO – Instituto de Ecología.

Guzmán, G. 2001. Medicinal fungi in Mexico: traditions, myths and knowledge. *International Journal of Medicinal Mushrooms,* 3: 95.

Hall, I. & Wang, Y. 2002. Truffles and other edible mycorrhizal mushrooms – some new crops for the southern hemisphere. *In* I.R. Hall, Y. Wang, A. Zambonelli & E. Danell, eds. *Edible ectomycorrhizal mushrooms and their cultivation.* Proceedings of the second international conference on edible ectomycorrhizal mushrooms. July 2001, Christchurch. CD-ROM. Christchurch, New Zealand Institute for Crop and Food Research Limited.

Hall, I., Zambonelli, A. & Primavera, E. 1998. Ectomycorrhizal fungi with edible fruiting bodies 3. *Tuber magnatum,* Tuberaceae. *Economic Botany,* 52(2): 192–200.

Hall, I.R., Buchanan, P.K., Wang, Y. & Cole, A.L.J. 1998a. *Edible and poisonous mushrooms: an introduction.* Christchurch, New Zealand Institute for Crop and Food Research Limited. 248 pp.

Hall, I.R., Lyon, A.J.E., Wang, Y. & Sinclair, L. 1998b. Ectomycorrhizal fungi with edible fruiting bodies 2. *Boletus edulis. Economic Botany,* 52(1): 44–56.

Hall, I.R., Stephenson, S., Buchanan, P., Wang, Y. & Cole, A.L.J. 2003. *Edible and poisonous mushrooms of the world.* Portland, Oregon, USA, Timber Press.

Halling, R.E. 1996. Recommendations for collecting mushrooms. *In* M.N. Alexiades. *Selected guidelines for ethnobotanical research: a field manual,* pp. 136–141. New York, USA, New York Botanical Garden.

Härkönen, M. 1988. Training people to collect and sell natural products in Finland. *Acta Botanica Fennica,* 136: 15–18.

Härkönen, M. 1995. An ethnomycological approach to Tanzanian species of Amanita. *Acta Universitas Uppsala. Symb. Bot. Ups.,* 30(3): 145–151.

Härkönen, M. 1998. Uses of mushrooms by Finns and Karelians. *International Journal of Circumpolar Health,* 40: 40–55.

Härkönen, M. 2000. The fabulous forests of Southern China as a cooperative field of exploration. *Universitas Helsingiensis,* 19 [XIX]: 20–22.

Härkönen, M. 2002. Mushroom collecting in Tanzania and Hunan (southern China): inherited wisdom and folklore of two different cultures. *In* R. Watling, J.C. Frankland, A.M. Ainsworth, S. Isaac & C.H. Robinson, eds. *Tropical mycology,* Vol. 1 *Macromycetes,* pp. 149–165. Wallingford, UK, CAB International.

Härkönen, M. & Järvinen, I. 1993. Evaluation of courses for mushroom advisors in Finland. *Aquilo, Ser. Botanica,* 31: 93–97.

Härkönen, M., Niemelä, T. and Mwasumbi, L. 2003. *Tanzanian mushrooms. Edible, harmful and other fungi, Norrlinea 10.* Helsinki, Botanical Museum, Finnish Museum of Natural History. 200 pp.

Härkönen, M., Saarimäki, T. & Mwasumbi, L. 1994a. Edible and poisonous mushrooms of Tanzania. *The African Journal of Mycology and Biotechnology,* 2(2): 99–123.

Härkönen, M., Saarimäki, T. & Mwasumbi, L. 1994b. Tanzanian mushrooms and their uses. 4. Some reddish edible and poisonous *Amanita* species. *Karstenia,* 34: 47–60.

Härkönen, M., Saarimäki, T. & Mwasumbi, L. 1995. Edible mushrooms of Tanzania. *Karstenia,* 35 supplement: 92.

Härkönen, M., Saarimäki, T., Mwasumbi, L. & Niemela, T. 1993. Collection of the Tanzanian mushroom heritage as a form of developmental cooperation between the universities of Helsinki and Dar es Salaam. *Aquilo, Ser. Botanica,* 31: 99–105.

Harsh, N.S.K., Rai, B.K. & Ayachi, S.S. 1993. Forest fungi and tribal economy – a case study in Baiga tribe of Madhya Pradesh [India]. *Journal of Tropical Forestry,* 9: 270–279.

Harsh, N.S.K., Rai, B.K. & Soni, V.K. 1999. Some ethnomycological studies from Madhya Pradesh, India. *In* J. Singh & K.R. Aneja, eds. *From ethnomycology to fungal biotechnology,* pp. 19–31. New York, USA, Plenum Press.

Harsh, N.S.K., Tiwari, C.K. & Rai, B.K. 1996. Forest fungi in the aid of tribal women of Madhya Pradesh [India]. *Sustainable Forestry,* 1: 10–15.

Hazani, E., Taitelman, U. & Sasha, S.M. 1983. *Amanita verna* poisoning in Israel – a report of a rare case out of time and place. *Archives of Toxicology,* Supplement 6: 186–189.

He, X. 1991. *Verpa bohemica* – a seldom known and delicious edible fungus. *Zhongguo Shiyongjun (Edible Fungi of China: a bimonthly journal),* 10(6): 19.

Hedger, J. 1986. *Suillus luteus* on the Equator. *Bulletin of the British Mycological Society,* 20: 53–54.

Heim, R. 1964. Note succincte sur les champignons alimentaires des Gadsup (Nouvelle Guinée. *Cahiers du Pacifique,* 6: 121–132.

Heyne, K. 1927. *De Nuttige Planten van Nederlandsch Indie. 2e [The useful plants of the Dutch East Indies]* 3 vols. Batavia, Dutch East Indies. Departement van Landbouw, Nijverheid en Handel. 1953 pp.

Hettula, A. 1989. Mushrooms in ancient Greece and Rome. *Opuscula, Instituti Romani Finlandiae,* 4: 17–42.

Hobbs, C. 1995. *Medicinal mushrooms: an exploration of tradition, healing, & culture.* 2nd edition. Santa Cruz CA, USA. Botanica Press. 252 pp.

Huang, N. 1989. New method of increasing production on *Rhizopogon piceus* in the south of Fujian Province. *Zhongguo Shiyongjun (Edible Fungi of China: a bimonthly journal),* 8 (overall No. 39)(5): 8.

Huang, N. 1993. *Edible fungi cyclopedia.* Beijing, Agricultural Publishing House of China. 448 pp.

Imazeki, R., Otani, Y., Hongo, T., Izawa, M. & Mizuno, N. 1988. *Fungi of Japan.* Tokyo, Yama-kei.

Iordanov, D., Vanev, S.G. & Fakirova, V.I. 1978. *Gubite v Bulgariya: Opredelitel na nai-razprostranenite yadlivi i otrovni gubi [Fungi of Bulgaria: keys to the identification of the most widely distributed edible and poisonous fungi].* Sofiya, Izd-vo na Bulg. Akad. na Naukite.

Isiloglu, M. & Watling, R. 1992. Macromycetes of Mediterranean Turkey. *Edinburgh Journal of Botany,* 49(1): 99–121.

Ivancevic, B. 1997. Conservation of fungi in Yugoslavia. *In* C. Perini, ed. *Conservation of fungi,* pp. 51–56. Siena, Italy, Universita degli Studi di Siena.

Jacobson, K.M. 1996. Macrofungal ecology in the Namib desert: a fruitful or futile study? *McIlvainea,* 12 (2): 21–32.

Jalkanen, R. & Jalkanen, E. 1978. Studies on the effects of soil surface treatments on crop of false morel (*Gyromitra esculenta*) in spruce forests. *Karstenia,* 18 (supplement): 56–57.

Jones, E.B.G. & Lim, G. 1990. Edible mushrooms in Singapore and other southeast Asian countries. *Mycologist,* 4: 119–124.

Jones, E.B.G., Whalley, A.J.S. & Hywel-Jones, N.L. 1994. A fungus foray to Chiang Mai market in Northern Thailand. *Mycologist,* 8(2): 87–90.

Kalamees, K. & Silver, S. 1988. Fungal productivity of pine heaths in North-West Estonia. *Acta Botanica Fennica,* 136: 95–98.

Kalinowski, M. 1998. Non-wood forest products in Poland. *In* H.G. Lund, B. Pajari & M. Korhonen, eds. *Sustainable development of non-wood goods and benefits from boreal and cold temperate forests,* pp. 87–92. Proceedings of the International Workshop, Joensuu, Finland, 18–22 January 1998. EFI-Proceedings. 1998, No. 23.

Kalotas, A. 1997. Aboriginal knowledge and use of fungi. In *Fungi of Australia.* Vol. 1B. *Introduction – Fungi in the environment,* pp. 269–295. Canberra, Australian Biological Resources Study.

Kasik, G. & Ozturk, C. 1995. Some edible, poisonous and non-edible macrofungi in Aksaray [in Turkish]. *Turkish Journal of Botany,* 19: 401–403.

Katende, A.B., Segawa, P. & Birnie, A. 1999. *Wild food plants and mushrooms of Uganda.* Nairobi, Kenya, Regional Land Management Unit, Swedish International Development Cooperation Agency. 490 pp.

Kaul, T.N. 1993. Conservation of mushroom resources in India. *Mushroom Research,* 2: 11–18.

Kawagoe, S. 1924. The market fungi of Japan. *Transactions of the British Mycological Society,* 10: 201–206.

Keewaydinoquay. 1998. *Puhpohwee for the people: a narrative account of some uses of fungi among the Ahnishinaabeg.* Dekalb, Illinois, USA, LEPS Press, Northern Illinois University. 67 pp.

Kiger, C.J. 1959. Etude de la composition chimique et de la valeur alimentaire de 57 espèces de champignons supérieurs. *Revue de Mycologie,* 24: 161–170.

Kim, Y.S. & Kim, S.S. 1990. *Illustrated Korean mushrooms.* Seoul, Yupoong Publ. 390 pp.

Kirk, P.M., Cannon, P.F., David, J.C. & Stalpers, J.A. 2001. *Dictionary of the fungi.* 9th edition. Wallingford, UK, CAB International. 655 pp.

Koistinen, R. 1978. The commercial mushroom yield in Northern Finland in 1976. *Karstenia,* 18 (supplement): 108–111.

Koo, C.D. & Bilek, E.M. 1998. Financial analysis of vegetation control for sustainable production of Songyi (*Tricholoma matsutake*) in Korea. *Journal of Korean Forest Society,* 87(4): 519–527.

Kovalenko, A. 1997. The present state of the conservation of fungi in Russia. *In* C. Perini, ed. *Conservation of fungi,* pp. 65–68. Siena, Italy, Universita degli Studi di Siena.

Kreula, M., Saarivirta, M. & Karanko, S.L. 1976. On the composition of nutrients in wild and cultivated mushrooms. *Karstenia,* 16: 10–14.

Kroeger, P. 1985. Mushrooms imported by Germany. *Mycena News,* 35: 3.

Kujala, M. 1988. Ten years of inquiries on the berry and mushroom yields in Finland, 1977–1986. *Acta Botanica Fennica,* 136: 11–13.

Kytovuori, I. 1989. The *Tricholoma caligatum* group in Europe and North Africa. *Karstenia,* 28: 65–77.

Lampe, K.F. & Ammirati, J.F. 1990. Human poisoning by mushrooms in the genus *Cortinarius. McIlvainea,* 9(2): 12–25.

Lau, O. 1982. Methods of chemical analysis of mushrooms. *In* S.T. Chang & T.H. Quimio, eds. *Tropical mushrooms. Biological nature and cultivation methods,* pp. 87–116. Hong Kong, Chinese University Press.

Lawrynowicz, L. 1997. Conservation of fungi in Poland. *In* C. Perini, ed. *Conservation of fungi,* pp. 25–30. Siena, Italy, Universita degli Studi di Siena.

Legg, A. 1991. Your top twenty fungi – the final list. *Mycologist,* 4: 23–24.

Leon-Guzman, M.F., Silva, I. & Lopez, M.G. 1997. Proximate chemical composition, free amino acid contents and free fatty acid contents of some edible mushrooms from Queretaro, Mexico. *Journal of Agricultural and Food Chemistry,* 45: 4329–4332.

Li, Z.P. 1994. Comparison of medicinal effect between wild *Ganoderma applanatum* and cultivated *Ganoderma lucidum. Zhongguo Shiyongjun (Edible Fungi of China: a bimonthly journal),* 13(2): 8–9.

Lincoff, G. 2002. There are only a dozen basic groups. *Mushroom, the Journal of Wild Mushrooming,* 20: 9–15.

Lincoff, G. & Mitchel, D.H. 1977. *Toxic and hallucinogenic mushroom poisoning. A handbook for physicians and mushroom hunters.* New York, USA, Van Nostrand Reinhold Company. 267 pp.

Liu, P.G. 1990. Investigation of the edible mushroom resources of Mt. Daqing of Inner Mongolia. *Zhongguo Shiyongjun [Edible Fungi of China],* 9: 26–27.

Liu, W.P. & Yang, H.R. 1982. An investigation of mushroom poisoning in Ninghua county during the last 20 years. *Chinese Journal of Preventative Medicine,* 16: 226–228.

Locquin, M. 1954. Une chanterelle comestible de la Côte d'Ivoire: *Hygrophoropsis mangenotii* sp. nov. *J. Agric. Bot. Trop. Appl.,* 1: 359–361.

Logemann, H., Argueta, J., Guzman, G., Montoya-Bello, L., Bandala-Munoz, V.M.

& **de Leon-Chocooj, R.** 1987. Lethal poisoning by mushrooms in Guatemala. *Revista Mexicana de Micología,* 3: 211–216.

Lopez, G.A., Cruz, J.M.M. & Zamora-Martinez, M.C. 1992. Evaluación de la produccion de hongos comestibles silvestres en San Juan Tetla, Puebla. Ciclo 1992. In *Reunion Científica Forestal y Agropecuaria,* pp. 182–191. Coyocan, Mexico.

Lowore, J. & Boa, E. 2001. *Bowa markets: local practices and indigenous knowledge of wild edible fungi.* Egham, UK, CABI Bioscience.

Lowore, J., Munthali, C. & Boa, E. 2002. *Bowa marketing channels and indigenous knowledge in Mzimba District, Malawi.* Egham, UK, CABI Bioscience.

Lowy, B. 1971. New records of mushroom stones from Guatemala. *Mycologia, 63*: 983–993.

Lowy, B. 1974. *Amanita muscaria* and the Thunderbolt legend in Guatemala and Mexico. *Mycologia,* 66: 189–191.

Lund, H.G., Pajari, B. & Korhonen, M. 1998. *Sustainable development of non-wood goods and benefits from boreal and cold temperate forests.* Proceedings of the International Workshop, Joensuu, Finland, 18–22 January 1998. EFI Proceedings No. 23. Joenssu, Finland, European Forest Institute.

Malencon, G. & Bertault, R. 1975. *Flore des champignons supérieurs du Maroc.* 2 volumes. Rabat, Faculté des Sciences. Vol. 2. 539 pp.

Malyi, L.P. 1987. Resources of edible fungi in Belorussia [Belarus] and the possibility of their utilization. *Rastitelo'nye Resursy,* 23(4): 532–536.

Mao, X.L. 1998. *Economic fungi of China (in Chinese).* Beijing. 762 pp.

Mao, X.L. 2000. *The macrofungi of China.* Beijing, Henan Science and Technology Press. 719 pp. (available at www.hceis.com).

Mao, X.L. & Jiang, C.P. 1992. *Economic macrofungi of Tibet.* Beijing, Beijing Science and Technology Publishing House. 651 pp.

Mapes, C., Guzmán, G. & Cabellero, J. 1981. Elements of the Purepecha mycological classification. *Journal of Ethnobiology,* 1(2): 231–237.

Markham, P. 1998. Fungal food in Fiji: a suspiciously familiar story. *Mycologist,* 12(1): 23–25.

Marles, R.J., Clavelle, C., Monteleone, L., Tays, N. & Burns, D. 2000. *Aboriginal plant use in Canada's northwest boreal forest.* Vancouver, Canada, University of British Columbia.

Martínez, A., Oria de Rueda, J.A. & Martínez, P. 1997. *Estudio sobre la potencialidad de los diferentes usos del bosque para la creación de empleo y actividad económica en el medio rural de Castilla León.* Universidad de Report for the Junta de Castilla y León y Fondo Social Europeo. 348 pp.

Martínez-Carrera, D., Aguilar, A., Martinez, W., Morales, P., Sobal, M., Bonilla, M. & Larque-Saavedra, A. 1998. A sustainable model for rural production of edible mushrooms in Mexico. *Micologia Neotropica Aplicada,* 11: 77–96.

Martínez-Carrera, D., Bonilla, M., Martinez, W., Sobal, M., Aguilar, A. & Pellicer-Gonzalez, E. 2001. Characterization and cultivation of wild *Agaricus* species in Mexico. *Micologia Aplicada International,* 13: 9–24.

Martínez-Carrera, D., Vergara, F., Juarez, S., Aguilar, A., Sobal, M., Martinez, W. & Larque-Saavedra, A. 1996. Simple technology for canning cultivated edible mushrooms in rural conditions in Mexico. *Micologia Neotropical Applicada,* 9: 15–27.

Martínez-Carrera, D., Morales, P., Pellicer-González, E., León, H., Aguilar, A., Ramírez, P., Ortega, P., Largo, A., Bonilla, M. & Gómez, M. 2002. Studies on the traditional management and processing of matsutake mushrooms in Oaxaca, Mexico. *Micologia Applicada International,* 14: 25–42.

Martinez-de-Aragón, J., Florit, E. & Colinas, C. 1998. Producción de setas micorrícicas y comestibles en la comarca del Solsones en 1997. In *III Forum de Política Forestal,* pp. 322–328. Solsona (Lleida), Centre Tecnologic Forestal de Catalunya.

Martins, A., Baptista, P., Sousa, M.J., Meireles, T. & Pais, M.S. 2002. Edible mycorrhizal fungi with *Castanea sativa* trees in the north-east of Portugal. *In* I.R. Hall, Y. Wang, A. Zambonelli & E. Danell, eds. *Edible ectomycorrhizal mushrooms and their cultivation.* Proceedings of the second international conference on edible ectomycorrhizal mushrooms. July 2001, Christchurch. CD-ROM. Christchurch, New Zealand Institute for Crop and Food Research Limited.

Mata, G. 1987. Introducción a la etnomicología maya de Yucatán, el conocimiento de los hongos en Pixoy, Valladolid. *Revista Mexicana Micología,* 3: 175–187.

Mata, M. 2003 *Macrohongos de Costa Rica.* Volume 1. Santo Domingo de Heredia, Costa Rica: INBIO. 256 pp.

Mata, M., Halling, R. & Mueller, G.M. 2003. *Macrohongos de Costa Rica.* Volume 2. Santo Domingo de Heredia, Costa Rica: INBIO. 256 pp.

Mata-Hidalgo, M. 1999. *Macrohongos de Costa Rica.* San José, Costa Rica, Instituto Nacional de Biodiversidad.

Matsuk, T. 2000. Picking through Russia's field of autumn. *The Russia Journal,* 36 16 Sept.

McKenzie, E.H.C. 1997. *Collect fungi on stamps.* London, Stanley Gibbons Ltd.

McLain, R.J., Christensen, H.H. & Shannon, M.A. 1998. When the amateurs are experts: amateur mycologists and wild mushroom politics in the Pacific Northwest. *USA Society for Natural Resources,* 11: 615–626.

Meijer, A.A.R. 2001. Mycological work in the Brazilian state of Paraná. *Nova Hedwigia,* 72: 105–159.

Mendoza, J.M. 1938. Philippines mushrooms. *Philippine Journal of Science,* 65: 1–128 (and 79 plates.

Mildh, U. 1978. The organization for collecting forest mushrooms in Finland. *Karstenia,* 18 (suppl.): 106–107.

Minter, D., Cannon, P.F. & Peredo, H.L. 1987. South America species of *Cyttaria* (a remarkable and beautiful group of edible ascomycetes). *Mycologist* 1: 7–11.

Molina, R., O'Dell, T., Luoma, D., Amaranthus, M., Castellano, M. & Russell, K. 1993. *Biology, ecology, and social aspects of wild edible mushrooms in the forests of the Pacific northwest: a preface to managing commercial harvest.* Portland, Oregon, USA, US Department of Agriculture, Forest Service, Pacific North-West Research Station. 42 pp.

Molina, R., Vance, N., Weigand, J., Pilz, D. & Amaranthus, M. 1997. Special forest products: integrating social, economic and biological considerations into ecosystem management. *In* K. Kohn & J. Franklin, eds. *Creating a forestry for the 21st century. The science of ecosystem management,* pp. 315–336. Washington, DC, Island Press.

Montoya-Esquivel, A. 1998. Ethnomycology of Tlaxcala, Mexico. *McIlvainea,* 13(2): 6–12.

Montoya-Esquivel, A., Estrada-Torres, A., Kong, A. & Juarez-Sanchez, L. 2001. Commercialization of wild mushrooms during market days of Tlaxcala, Mexico. *Micologia Aplicada International,* 13: 31–40.

Moore, A. 1996. Meeting Asian pickers (available at www.matsiman.com).

Moreno-Arroyo, B., Recio, J.M., Gomez, J. & Pulido, E. 2001. *Tuber oligospermum* from Morocco. *Mycologist,* 15: 41–42.

Moreno-Fuentes, A., Cifuentes, J., Bye, R. & Valenzuela, R. 1996. Kute-mo'ko-a: an edible fungus of the Raramuri Indians of Mexico. *Revista Mexicana de Micología,* 12: 31–39.

Morris, B. 1984a. Macrofungi of Malawi: some ethnobotanical notes. *Bulletin of the British Mycological Society,* 18: 48–57.

Morris, B. 1984b. The pragmatics of folk classification. *Journal of Ethnobiology,* 4(1): 45–60.

Morris, B. 1987. *Common mushrooms of Malawi.* Oslo, Fungiflora. 108 pp.

Morris, B. 1992. Mushrooms: for medicine, magic and munching. *Nyala,* 16(1): 1–8.

Morris, B. 1994. Bowa: Ethnomycological notes on the macrofungi of Malawi. *In* J.H. Seyani & A.C. Chikuni, eds. *Proceedings of the XIIIth Plenary meeting of AETFAT,* Vol.

1, pp. 635–647. Zomba, Malawi, 2–11 April 1991. Zomba, Malawi, National Herbarium and Botanic Gardens of Malawi.

Mshigeni, K.E. & Chang, S.T., eds. 2000. *A guide to successful mushroom farming: with emphasis on technologies appropriate and accessible to Africa's rural and peri-urban communities.* UNDP/UNOPS regional project RAF/99/021. Windhoek, University of Namibia. 34 pp.

Mushroom, the Journal of Wild Mushrooming. 2002. Go ahead and eat them: Matsutake are not endangered. 20: 7–8.

Namgyel, P. 2000. The story of Buddha mushroom. *Tricholoma matsutake*. Unpublished manuscript, Thimpu, Bhutan. 14 pp.

Niemela, T. & Uotila, P. 1977. Lignicolous macrofungi from Turkey and Iran. *Karstenia,* 17: 33–39.

Nieves-Rivera, A.M. 2001. Origin of mycophagy in the West Indies. *Inoculum: newsletter of the Mycological Society of America,* 52(2): 1–3.

Novellino, D. 1999. *The ominous switch: from indigenous forest management to conservation – the case of the Batak on Palawan Island, Philippines.* Copenhagen, IWGIA.

Obodai, M. & Apetorgbor, M. 2001. *An ethnobotanical study of mushroom germplasm and its domestication in the Bia Biosphere Reserve of Ghana.* Report presented to UNESCO through Environmental Protection Agency of Ghana, Accra.

Ohenoja, E. 1978. Mushrooms and mushroom yields in fertilised forests. *Acta Botanica Fennica,* 15: 38–46.

Ollikainen, T. 1998. Belarus forestry strategic plan and the non-wood forest products. *In* H.G. Lund, B. Pajari & M. Korhonen, eds. *Sustainable development of non-wood goods and benefits from boreal and cold temperate forests,* pp. 159–165. Proceedings of the International Workshop, Joensuu, Finland, 18–22 January 1998. EFI-Proceedings. 1998, No. 23.

Oso, B. 1975. Mushrooms and the Yoruba people of Nigeria. *Mycologia,* 67(2): 311–319.

Oso, B. 1977. Mushrooms in Yoruba mythology and medicinal practices. *Economic Botany,* 31: 367–371.

Paal, T. 1998. Utilisation and research of non-wood products in the former Soviet Union. *In* H.G. Lund, B. Pajari & M. Korhonen, eds. *Sustainable development of non-wood goods and benefits from boreal and cold temperate forests,* pp. 119–124. Proceedings of the International Workshop, Joensuu, Finland, 18–22 January 1998. EFI-Proceedings. 1998, No. 23.

Paal, T. 1999. Wild berry and mushrooms resources in Estonia and their exploitation. *Metsanduslikud Uurimused,* 31: 131–140.

Paal, T. & Saastamoinen, O. 1998. Non-wood plant products in Estonian forests. *In* H.G. Lund, B. Pajari & M. Korhonen, eds. *Sustainable development of non-wood goods and benefits from boreal and cold temperate forests,* pp. 109–117. Proceedings of the International Workshop, Joensuu, Finland, 18–22 January 1998. EFI-Proceedings. 1998, No. 23.

Pakistan Economist. 2001. Bright prospects for mushroom exports (morels).

Parent, G. & Thoen, D. 1977. Food value of edible mushrooms from Upper Shaba region. *Economic Botany,* 31: 436–445.

Park, W.H. & Lee, H.D. 1999. *Korean mushrooms.* Seoul, Kyo-Hak-Sa.

Pauli, G. 1998. *Qingyuan: the mushroom capital of the world* (available at www.zeri.org/news/1998/august/aug_chin.htm).

Pauli, G. 1999. *Sustainable development in the Amazon forest* (available at www.zeri.org).

Peerally, A. 1979. *Tricholoma spectabilis,* an excellent giant edible mushroom from Mauritius. *In* J. Delmas, ed. *Mushroom science X,* pp. 817–828. Proceedings of the Tenth International Congress on the Science and Cultivation of Edible Fungi. France, 1978.

Pegler, D.N. & Piearce, G.D. 1980. The edible mushrooms of Zambia. *Kew Bulletin,* 35: 475–491.

Pegler, D.N. & Vanhaecke, M. 1994. Termitomyces of southeast Asia. *Kew Bulletin,* 49: 717–736.

Pekkarinen, M. & Maliranta, H. 1978. Preliminary study of the consumption of mushrooms in Finland. *Karstenia,* 18 (suppl.): 47–48.

Perini, C. ed. 1998. *Conservation of fungi in Europe.* Proceedings of the 4th meeting of the European Council for the Conservation of Fungi. Vipiteno (Sterzing, Italy), 9–14 September 1997. Siena, Italy, Universita degli Studi de Siena. 159 pp.

Phillips, R., Shearer, L., Reid, D. & Rayner, R. 1983. *Mushrooms and other fungi of Great Britain and Europe.* London, Pan. 288 pp.

Piearce, G.D. 1981. Zambian mushrooms – customs and folklore. *Bulletin of the British Mycological Society,* 15(2): 139–142.

Piearce, G.D. 1985. Livingstone and fungi in tropical Africa. *Bulletin of the British Mycological Society,* 19(1): 39–50.

Pilát, A. 1951. *Mushrooms.* London, Spring Books. 120 pp.

Pilz, D. & Molina, R. 2002. Commercial harvest of edible mushrooms from the forests of the Pacific Northwest United States: issues, management and monitoring for sustainability. *Forest Ecology and Management,* 155: 3–16.

Pilz, D., Smith, J., Amaranthus, M.P., Alexander, S., Molina, R. & Luoma, D. 1999. Mushrooms and timber. Managing commercial harvesting in the Oregon Cascades. *Journal of Forestry* 97: 4–11.

Plum, P.M. 1998. Denmark: non-wood forestry in a densely populated temperate country. *In* H.G. Lund, B. Pajari & M. Korhonen, eds. *Sustainable development of non-wood goods and benefits from boreal and cold temperate forests*, pp. 125–130. Proceedings of the International Workshop, Joensuu, Finland, 18–22 January 1998. EFI-Proceedings. 1998, No. 23.

Pop, A. 1997. A short report on fungi conservation in Romania. *In* C. Perini, ed. *Conservation of fungi*, p. 141. Siena, Italy, Universita degli Studi di Siena.

Prance, G. 1984. The use of edible fungi by Amazonian Indians. *Advances in Economic Botany,* 1: 127–139.

Prasad P., Chauhan, K., Kandari, L.S., Maikhuri, R.K., Purohit, A., Bhatt, R.P. & Rao, K.S. 2002) *Morchella esculenta*: need for scientific intervention for its cultivation in central Himalaya. *Current Science,* 82: 1098–1022.

Purkayastha, R.P. & Chandra, A. 1985. *Manual of edible mushrooms.* New Delhi, Today and Tomorrow's Printers and Publishers.

Rai, B.K., Ayachi, S.S. & Rai, A. 1993. A note on ethno-myco-medicines from Central India. *Mycologist,* 7: 192–193.

Rammeloo, J. 1994. The contributions of the national botanic garden of Belgium to the mycology of Africa. *In* J.H. Seyani & A.C. Chikuni, eds. *Proceedings of the XIIIth Plenary meeting of AETFAT*, Zomba, Malawi, 2–11 April 1991, Vol. 1, pp. 671–685. Zomba, Malawi, National Herbarium and Botanic Gardens of Malawi.

Rammeloo, J. & Walleyn, R. 1993. The edible fungi of Africa south of the Sahara: a literature survey. *Scripta Botanica Belgica,* 5: 1–62.

Rautavaara, T. 1947. *Suomen sienisato. Summary: Studies on the mushroom crop in Finland and its utilisation.* ?Werner Spederstrom Osakeyhtio, Forssan Kirjapaino Oy.

Redhead, S.A. 1997. The pine mushroom industry in Canada and the United States: why it exists and where it is going. *In* I.H. Chapela & M.E. Palm eds. *Mycology in sustainable development: expanding concepts,* pp. 15–39. Boon, North Carolina, Parkway Publishers.

Remotti, C.D. & Colan, J.A. 1990. Identification of wild edible fungi in Dantas Forest, Huanuco. *Revista Forestal del Peru,* 17: 21–37.

Reshetnikov, S.V., Wasser, S.P. & Tan, K.K. 2001. Higher basidiomycota as a source of Antitumour and immunostimulating polysaccharides. A review. *International Journal of Medicinal Mushrooms,* 3: 361–394.

Reygadas, F., Zamora-Martinez, M. & Cifuentes, J. 1995. Conocimiento sobre los hongos silvestres comestibles en las comunidades de Ajusco y Topilejo D.F. *Revista Mexicana de Micologia* 11: 85–108.

Reyna, S., Rodriguez-Barreal, J., Folch, L., Perez-Badia, R. ,Garcia, S. & Jimenez, E. 2002. Truffle silviculture in Mediterranean forests. *In* I.R. Hall, Y. Wang, A. Zambonelli & E. Danell, eds. *Edible ectomycorrhizal mushrooms and their cultivation.* Proceedings of the second international conference on edible ectomycorrhizal mushrooms. July 2001, Christchurch. CD-ROM. Christchurch, New Zealand Institute for Crop and Food Research Limited.

Richards, A. 1939. *Land, Labour and Diet in Northern Rhodesia. An economic study of the Bemba tribe.* London, UK, Oxford University Press.

Richards, R.T. & Creasy, M. 1996. Ethnic diversity, resource values and ecosystem management: matsutake mushroom harvesting in the Klamath bioregion. *Society and Natural Resources,* 9: 359–374.

Richardson, D.H.S. 1988. Medicinal and other economic aspects of lichens. *In* M. Galun, ed. *CRC handbook on lichenology,* Volume 3. pp. 93–108. Baton Rouge, CRC.

Richardson, D.H.S. 1991. Lichens and man. *In* D.L. Hawksworth, ed. *Frontiers in mycology,* pp. 187–210. Wallingford, CAB International.

Riedlinger, T.J., ed. 1990. *The sacred mushroom seeker. Essays for R. Gordon Wasson.* Portland, Oregon, Dioscorides Press. 283 pp.

Rifai, M. 1989. van Overeem's unpublished icones of Indonesian edible fungi. *In* J.S. Siemonsma & N. Wuligarni-Soetjipto, eds. *Plant resources of South-East Asia,* pp. 297–298. Proceedings of the first PROSEA International Symposium, 22–35 May 1989, Jakarta, Indonesia.Wageningen, the Netherlands, PUDOC/PROSEA.

Rijsoort, J.V. & Pikun, H. 2000. *International Seminar on Non-Timber Forest Product – China Yunnan, Laos, Vietnam.* Simao City, Yunnan, PR China, Yunnan University Press. 187 pp.

Rodríguez, J.A., Llamas-Frade, B., Terrón-Alfonso, A., Sánchez-Rodríguez, J.A.,García-Prieto, O., Arrojo-Martín, E. & Jarauta, T.P. 1999. *Guía de Hongos de la Península Ibérica.* 3rd edition. León, Celarayn.

Rojas, C. & Mansur, E. 1995. Ecuador: informaciones generales sobre productos non madereros en Ecuador. In *Memoria, consulta de expertos sobre productos forestales no madereros para America Latina y el Caribe,* pp. 208–223. Serie Forestal #1. Santiago, Chile, FAO Regional Office for Latin America and the Caribbean.

Rotheroe, M. 1998. Wild fungi and the controversy over collecting for the pot. *British Wildlife,* 9(6): 349–355.

Rutkauskas, A. 1998. Non-wood resources and their utilisation in Lithuania. *In* H.G. Lund, B. Pajari & M. Korhonen, eds. *Sustainable development of non-wood goods and benefits from boreal and cold temperate forests,* pp. 93–101. Proceedings of the International Workshop, Joensuu, Finland, 18–22 January 1998. EFI-Proceedings. 1998, No. 23.

Ryvarden, L., Piearce, G.D. & Masuka, A. 1994. *An introduction to the larger fungi of South Central Africa.* Oslo, Norway, Fungiflora.

Saar, M. 1991. Fungi in Khanty folk medicine. *Journal of Ethnopharmacology,* 31: 175–179.

Saastamoinen, O. 1999. Forest policies, access rights and non-wood forest products in northern Europe. *Unasylva,* 50: 20–26.

Saastamoinen, O.,Kangas, J., Naskali, A. & Salo, K. 1998. Non-wood forest products in Finland: statistics, expert estimates and recent development. *In* H.G. Lund, B. Pajari & M. Korhonen, eds. *Sustainable development of non-wood goods and benefits from boreal and cold temperate forests,* pp. 131–146. Proceedings of the International Workshop, Joensuu, Finland, 18–22 January 1998. EFI-Proceedings. 1998, No. 23.

Sabra, A. & Walter, S. 2001. *Non-wood forest products in the Near East: a regional and national overview.* Working paper FOPW/01/2. Rome, FAO. 120 pp.

Saenz, J.A., Lizano, A.V.M. & Nassar, M.C. 1983. Edible, poisonous and hallucinatory fungi in Costa Rica. *Revista de Biologia Tropical,* 31: 201–207.

Salo, K. 1999. Principles and design of a prognosis system for an annual forecast of non-wood forest products. *In* A. Niskanen & N. Demidova, eds. *Research approaches to support non-wood forest products sector development: case of Arkhangelsk Region, Russia,* pp. 35–44. European Forest Institute Proceedings No. 29. Joensuu, EFI.

Sanon, K.B., Ba, A.M. & Dexheimer, J. 1997. Mycorrhizal status of some fungi fruiting beneath indigenous trees in Burkina Faso. *Forest Ecology and Management,* 98: 61–69.

Saremi, H., Ammarellou, A. & Mohammadi, J. 2002. Morphological and ecological evaluation of truffles in Iran. *In* I.R. Hall, Y. Wang, A. Zambonelli & E. Danell, eds. *Proceedings of the second international conference on edible mycorrhizal mushrooms.* July 2001, Christchurch. CD-ROM. Christchurch, New Zealand Institute for Crop and Food Research Limited.

Sarkar, B.B., Chakraborty, D.K., & Bhattacharjee, A. 1988. Wild edible mushroom flora of Tripura. *Indian Agriculturist,* 32: 139–143.

SCBD. 2001. *Sustainable management of non-timber forest resources.* CBD Technical Series No. 6. Montreal, Secretariat of the Convention on Biological Diversity (available at www.biodiv.org). 30 pp.

Schlosser, W.E. & Blatner, K.A. 1995. The wild edible mushroom industry of Washington, Oregon and Idaho, a 1992 survey. *Journal of Forestry,* 93: 31–36.

Schmeda-Hirschmann, G., Razmilic, I., Reyes, S., Gutierrez, M.I. & Loyola, J.I. 1999a. Biological activity and food analysis of *Cyttaria* spp. (Discomycetes). *Economic Botany,* 53(1): 30–40.

Schmeda-Hirschmann, G., Razmilic, I., Gutierrez, M.I. & Loyola, J.I. 1999b. Proximate composition and biological activity of food plants gathered by Chilean Amerindians. *Economic Botany,* 53(2): 177–187.

Schultes, R.G. 1940. Teonancatl: the narcotic mushroom of the Aztecs. *American Anthropologist,* XLII: 429–443.

Sergeeva, M. 2000. *Fungi. 250 species of edible, poisonous and medicinal fungi.* Moscow, Culture and Traditions. 263 pp.

Shackleton, S.E., Shackleton, C.M., Netshiluvhi, T.R., Geach, B.S., Ballance, A. & Fairbanks, D.H.K. 2002. Use patterns and value of savanna resources in three rural villages in South Africa. *Economic Botany,* 56(2): 130–146.

Sharda, R.M., Kaushal, S.C. & Negi, G.S. 1997. Edible fungi of Garhwal Himalayas. *Mushroom Journal,* 1997: 11–13.

Sharma, Y.K. & Doshi, A. 1996. Some studies on an edible wild fungus *Phellorinia inquinans*, in Rajasthan, India. *Mushroom Research,* 5: 51–53.

Shaw, D. 1984. *Microorganisms in Papua New Guinea.* Research Bulletin No.33, Department of Primary Industry, Port Moresby.

Siddiqi, N.A. 1998. Ethnobotany of non-timber forest products of Chittagong Hill Tracts. *In* R.L. Banik, M.K. Alam, S.J. Pei & A. Rastogi, eds. *Applied ethnobotany,* pp. 52–55. Chittagong, Bangladesh, Bangladesh Forest Research Institute.

Sillitoe, P. 1995. Ethnoscientific observations on entomology and mycology in the southern highlands of Papua New Guinea. *Science in New Guinea,* 21(1): 3–26.

Simmons, C., Henkel, T. & Bas, C. 2002. The genus *Amanita* in the Pakaraima mountains of Guyana. *Persoonia,* 17(4): 563–582.

Simons, D.M. 1971. The mushroom toxins. *Delaware Medical Journal,* 43(7): 177–187.

Singer, R. 1953. Four years of mycological work in southern South America. *Mycologia,* 45: 865–891.

Singh, S.K. & Rawat, G.S. 2000. Morel mushroom industry in India. *Plant Talk,* 21: 36–37.

Sisak, L. 1998. Importance of main non-wood forest products in the Czech Republic. *In* H.G. Lund, B. Pajari & M. Korhonen, eds. *Sustainable development of non-wood goods and benefits from boreal and cold temperate forests,* pp. 79–85. Proceedings of the

International Workshop, Joensuu, Finland, 18–22 January 1998. EFI-Proceedings. 1998, No. 23.

Sommer, R. 1995. Why I will continue to eat corn smut. *Natural History*, 19–22.

Sommerkamp, I. & Guzmán, G. 1990. Hongos de Guatemala. II Especies depositadas en el herbario de la Universidad de San Carlos de Guatemala. *Revista Mexicana de Micología*, 6: 179–197.

Stamets, P. 2000. *Growing gourmet and medicinal mushrooms.* 3rd edition. Berkeley, California, Ten Speed Press. 574 pp.

Suhardi. 2000. Treatment to develop mycorrhiza formation on dipterocarp seedlings. *In* E. Guhardia, M. Fatawi, M. Sutisna, T. Mori & S. Ohta, eds. *Rainforest ecosytems of East Kalimantan*, pp. 245–250. Ecological Studies Vol. 140. Tokyo, Springer.

Sun, W.S. & Xu, J.Y. 1999. Cultivation of edible fungi has become one of the backbone industries in rural economy of China. *Edible Fungi of China*, 18(2): 5–6.

Syed-Riaz, A.G. & Mahmood-Khan, S. 1999. Edible mushrooms from Azad Jammu and Kashmir. *Pakistan Journal of Phytopathology*, 11: 163–165.

Tagliavini, O. & Tagliavini, R. 2001. *Atlante dei funghi commestibiliti della Basilicata (Atlas of edible fungi from Basilicata.* Potenza, Consiglio Regionale della Basilicata. 342 pp.

Taylor, F.W., Thamage, D.M., Baker, N., Roth-Bejerano, N. & Kagan-Zur, V. 1995. Notes on the Kalahari desert truffle, *Terfezia pfeillii. Mycological Research*, 99: 874–878.

Tedder, S., Mitchell, D. & Farran, R. 2000. *Seeing the forest beneath the trees: the social and economic potential of non-timber forest products and services in the Queen Charlotte Islands/Haida Gwaii.* Mitchell Consulting and the BC Ministry of Forests. British Columbia.

Tedder, S., Mitchell, D. & Farran, R. 2002. *Property rights in the sustainable management of non-timber forest products.* Victoria, British Columbia, British Columbia, Ministry of Forests. 140 pp.

Testi, A. 1999. *Il Libor dei Funghi d'Italia.* Colognola ai Colli (VR), Demetra. 384 pp.

Thoen, D. 1993. Looking for ectomycorrhizal trees and ectomycorrhizal fungi in tropical Africa. *In* S. Isaac, J.C. Frankland, R. Watling & A.J.S. Whalley, eds. *Aspects of tropical mycology*, pp. 193–205. Cambridge, Cambridge University Press.

Thoen, D. & Ba, A.M. 1989. Ectomycorrhizas and putative ectomycorrhizal fungi of *Afzelia africana* and *Uapaca senegalensis* in southern Senegal. *New Phytologist*, 113: 549–559.

Thomson, B.P. 1954. Two studies in African nutrition. An urban and a rural community in Northern Rhodesia. *Rhodes-Livingstone Papers*, 24: 77– 86.

Tibiletti, E. & Zambonelli, A. 1999. *I Tartufi della Provincia di Forli-Cesena*. Bologna, Patron Editore. 178 pp.

Trappe, J.M. 1990. Use of truffles and false truffles around the world. *In* M. Bencivenga & B. Granetti, eds. Proceedings, Atti del Secondo Congresso Internazionale sul Tartufo. Spoleto 1988. pp. 19–30. Spoleto, Italy, Comunita Montana dei Monti Martini e del Serano.

Tu, G.L. 1987. Using bagasse and waste cotton as substrate for bag cultivation of *Pleurotus sajor-caju. Zhongguo Shiyongjun (Edible Fungi of China: a bimonthly journal)* 6 (overall No. 23)(1): 30–34.

Tuno, N. 2001. Mushroom utilization by the Majangir, an Ethiopian tribe. *The Mycologist*, 15: 78–79.

Turner, N.J., Kuhnlein, H.V. & Egger, K.N. 1987. The cottonwood mushroom (*Tricholoma populinum*): a food resource of the Interior Salish Indian peoples of British Columbia. *Canadian Journal of Botany*, 65: 921–927.

Uaciquete, A., Dai, M.d.L. & Motta, H. 1996. *Distribução, valor economico e uso sustentavel do cogumelo comestível em Moçambique* [Distribution, economic value and sustainable use of edible mushrooms in Mozambique]. Grupo de Trabalho Ambiental [Environmental Working Group]. Maputo, Mozambique.

Urbonas, V., Kalamees, K. & Lukin, V. 1974. *The list of the agaricales flora of the Baltic Republics (Lithuania, Latvia, Estonia) [in Russian].* Publisher not known.

Vachuska, P. & Vachuska, C. 2000. Mushroom poisonings in Russia; mushroom deaths in the Ukraine. *The Newsletter of the Wisconsin Mycological Society,* 17(3.

Vaidya, J.G. & Rabba, A.S. 1993. Fungi in folk medicine. *Mycologist,* 7: 131–133.

Vance, N.C. & Thomas, J. 1995. *Special forest products. Biodiversity meets the marketplace.* Portland, Oregon, US Department of Agriculture, Forest Service.

van der Westhuizen, G.C.A. & Eicker, A. 1994. *Field guide: mushrooms of southern Africa.* Cape Town, Struik Publishers (Pty) Ltd. 207 pp.

Vasil'eva, L.N. 1978. *Edible mushrooms of the Far East.* Vladivostock, Far Eastern Publishing House.

Verbecken, A., Walleyn, R., Sharp, C. & Buyck, B. 2000. Studies on tropical African *Lactarius* species 9. Records from Zimbabwe. *Syst. Geogr. Pl.,* 70: 181–215.

Verbecken, A. & Buyck, B. 2002. Diversity and ecology of tropical ectomycorrhizal fungi in Africa. *In* R. Watling, J.C. Frankland, A.M. Ainsworth, S. Isaac & C.H. Robinson, eds. *Tropical mycology* Vol. 1, pp. 11–24. Macromycetes. Wallingford, CAB International.

Vilkriste, L. 1998. NWFP resources and their future utilisation in Latvia. *In* H.G. Lund, B. Pajari & M. Korhonen, eds. *Sustainable development of non-wood goods and benefits from boreal and cold temperate forests,* pp. 103–108. Proceedings of the International Workshop, Joensuu, Finland, 18–22 January 1998. EFI-Proceedings. 1998, No. 23.

Villanueva, C. 1997. 'Huitlacoche' (*Ustilago maydis*) as a food in Mexico. *Micologia Neotropica Aplicada,* 10: 73–81.

Villarreal, L. & Guzmán, G. 1985. Producción de lost hongos comestibles silvestres en los bosques de México I. *Revista de la Sociedad Mexicana de Micología,* 1: 51–90.

Villarreal, L. & Guzmán, G. 1986a. Producción de lost hongos comestibles silvestres en los bosques de México II. *Biotica,* 11: 271–280.

Villarreal, L. & Guzmán, G. 1986b. Producción de lost hongos comestibles silvestres en los bosques de México III. *Revista de la Sociedad Mexicana de Micologia,* 2: 259–277.

Villarreal, L. & Perez-Moreno. J. 1989. Los hongos comestibles silvestres de Mexico, un enfoque integral. *Micologia Neotropica Aplicada,* 2: 77–114.

Vladyshevskiy, D.V., Laletin, A.P. & Vladyshevskiy, A.D. 2000. Role of wildlife and other non-wood forest products in food security in central Siberia. *Unasylva,* 51: 46–52.

Walleyn, R. & Rammeloo, J. 1994. The poisonous and useful fungi of Africa south of the Sahara: a literature survey. *Scripta Botanica Belgica,* 10: 1–56.

Walker, A. 1931. Champignon comestibles de la Basse-Ngounié (Gabon). *Revue Bot App et Agriculture Tropical,* 11: 240–247.

Wang, Y.C. 1987. Mycology in ancient China. *Mycologist,* 1: 59–61.

Wang, Y., Hall, I.R. & Evans, L.A. 1997. Ectomycorrhizal fungi with edible fruiting bodies. 1. *Tricholoma matsutake* and related fungi. *Economic-Botany,* 51(3): 311–327.

Wang, Y, Buchanan, P. & Hall, I. 2002. A list of edible ectomycorrhizal mushrooms. *In* I.R. Hall, Y. Wang, A. Zambonelli & E. Danell, eds. *Edible ectomycorrhizal mushrooms and their cultivation.* Proceedings of the second international conference on edible ectomycorrhizal mushrooms. July 2001. CD-ROM. Christchurch, New Zealand Institute for Crop and Food Research Limited.

Wasser, S.P. 1990. *Edible and poisonous mushrooms of the Carpathian Mountains.* 2nd edition? Uzhgorod, Ukraine, Karpaty Press. 205 pp.

Wasser, S.P. 1995. *Edible and poisonous mushrooms of Israel.* Tel-Aviv, Modan Press. 185 pp.

Wasser, S.P., Nevo, E., Sokolov, D., Reshetnikov, S. & Timor-Tismenetsky, M. 2000. Dietary supplements from medicinal mushrooms: diversity of types and variety of regulations. *International Journal of Medicinal Mushrooms,* 2: 1–19.

Wasser, S.P. & Weis, A.L. 1999a. General description of the most important medicinal higher basidiomycetes mushrooms. 1. *International Journal of Medicinal Mushrooms,* 1: 351–370.

Wasser, S.P. & Weis, A.L. 1999b. Medicinal properties of substances occurring in higher basidiomycetes mushrooms: current perspectives (review). *International Journal of Medicinal Mushrooms,* 1: 31–62.

Wasser, S.P. & Weis, A.L. 1999c. Therapeutic effects of substances occurring in higher basidiomycetes mushrooms: a modern perspective. *Critical Reviews in Immunology,* 19: 65–96.

Wasson, R.G. 1968. *Soma, divine mushroom of immortality.* The Hague, Mouton. 381 pp.

Wasson, V.P. & Wasson, R.G. 1957. *Mushrooms, Russia and history.* 2 vols. New York, Pantheon Books.

Weigand, J.F. 1998. *Management experiments for high-elevation agroforestry systems jointly producing matsutake mushrooms and high-quality timber in the Cascade range of southern Oregon. General Technical Report PNW-GTR-424.* Portland, Oregon, US Department of Agriculture, Forest Service, Pacific Northwest Research Station. 42 pp.

Wills, R.M. & Lipsey, R.G. 1999. *An economic strategy to develop non timber forest products and services in British Columbia.* British Colombia, Ministry of Forests.

Wilson, K., Cammack, D. & Shumba, F. 1989. Food provisioning amongst Mozambican refugees in Malawi. A study of aid, livelihood and development. A report prepared for the World Food Programme. Oxford University. Oxford, UK.

Winkler, D. 2000. *Sustainable development in the Tibetan areas of West Sichuan after the logging ban.* Unpublished presentation for 9th IATS Symposium, Leiden, Netherlands, 24–30 June 2000 (available at http://ourworld.cs.com/danwink/daniel_winkler_s_selected_publications.htm?f=fs).

Winkler, D. 2002. Forest use and implications of the 1998 logging ban in the Tibetan prefectures of Sichuan: Case study on forestry, reforestation and NTFP in Litang County, Ganzi TAP, China. *In* Z. Ziang, M. Centritto, S. Liu & S. Zhang, eds. *The ecological basis and sustainable management of forest resources.* Informatore Botanico Italiano 134 (Supplemento 2): [in press]

Xiang, Y.T. & Han, Z. 1987. Using sun-cured bed to increase temperature in the early spring for culturing straw mushroom (*Volvariella esculenta. Zhongguo Shiyongjun (Edible Fungi of China: a bimonthly journal)* 6 (overall No. 23)(1): 16–17.

Yang, Z. 1990. A delicious tropical mushroom – *Termitomyces heimii* occurring in Yunnan, China. *Zhongguo Shiyongjun (Edible Fungi of China: a bimonthly journal),* 9(4): 28–30.

Yang, Z. 1992. *Polyozellus multiplex* – a rare edible fungus. *Zhongguo Shiyongjun (Edible Fungi of China: a bimonthly journal),* 11(2): 1–4.

Yang, Z.L. & Yang, C. 1992. Recognition of *Hypsizygus marmoreus* and its cultivation. *Zhongguo Shiyongjun (Edible Fungi of China: a bimonthly journal),* 11(5): 19–20.

Yeh, E. 2000. Forest claims, conflicts and commodification: the political ecology of Tibetan mushroom-harvesting villages in Yunnan Province, China. *China Quarterly,* 161: 225–278.

Yilmaz, F., Oder, N. & Isiloglu, M. 1997. The macrofungi of the Soma (Manisa) and Savastepe (Balikesir) districts. *Turkish Journal of Botany,* 21(4): 221–230.

Ying, J.Z., Mao, X.L., Ma, Q.M., Zong, Y.C. & Wen, H.A. 1987. *Icones of medicinal fungi from China.* Beijing, Science Press. 575 pp.

Ying, J.Z., Zhao, J.B., Mao, X.L., Ma, Q.M., Zhao, L.W. & Zong, Y.C. 1988. *Edible mushrooms [of China].* Beijing, Science Publishing House.

Yokoyama, K. 1975. Ainu names and uses for fungi, lichens and mosses. *Transactions of the Mycological Society, Japan,* 16: 183–9.

Yorou, S.N. & De Kesel, A. 2002. Connaissances ethnomycologiques des peuples Nagot du centre du Bénin (Afrique de l'Ouest). *In* E. Robbrecht, J. Degreef & I. Friis, eds. *Plant systematics and phytogeography for the understanding of African biodiversity.* Proceedings of the XVith AETFAT Congresss 2000, Meise, National Botanic Garden of Belgium. *Syst. Geogr. Pl.,* 71: 627–637.

Yorou, S.N., De Kesel, A., Sinsin, B. & Codjia, J.T.C. 2002. Diversité et productivité des champignons comestibles de la foret classée de Wari Maro (Benin). *In* E. Robbrecht, J.

Degreef & I. Friis, eds. *Plant systematics and phytogeography for the understanding of African biodiversity*. Proceedings of the XVith AETFAT Congress 2000, Meise, National Botanic Garden of Belgium. *Syst. Geogr. Pl.*, 71: 613–625.

Yun, W., Buchanan, P. & Hall, I. 2002. A list of edible ectomycorrhizal mushrooms. *In* I.R. Hall, Y. Wang, A. Zambonelli & E. Danell, eds. *Edible ectomycorrhizal mushrooms and their cultivation*. Proceedings of the second international conference on edible ectomycorrhizal mushrooms. July 2001, Christchurch. CD-ROM. Christchurch, New Zealand Institute for Crop and Food Research Limited.

Zakhary, J.W., Abo-Bakr, T.M., El-Mahdy, A.R. & El-Tabery, S.A.M. 1983. Chemical composition of wild mushrooms collected from Alexandria, Egypt. *Food Chemistry*, 11: 31–41.

Zaklina, M. 1998. *Edible mycorrhizal mushrooms in Serbia - problems with protection*. 2nd International Conference on Mycorrhiza, Uppsala, Sweden, 5–10 July 1998 (available at www.mycorrhiza.ag.utk.edu).

Zamora-Martinez, M.C., Alvardo, G. & Dominguez, J.M. 2000. *Hongos Silvestres Comestibles region de Zacualtipan, Hidalgo*. Pachuca, Hidalgo, Mexico, INIFAP CIR-CENTRO#.

Zamora-Martinez, M.C., Reygadas, G.F. & Cifuentes, J. 1994. *Hongos comestibles silvestres de la subcuenca Arroya El Zorrillo, Distrito Federal*. Coyoacan, DF Mexico, INIFAP.

Zang, M. 1984. Mushroom distribution and the diversity of habitats in Tibet, China. *McIlvainea*, 6(2): 15–20.

Zang, D.C. 1988a. *Collybia albuminosa* at Lianshan District. *Zhongguo Shiyongjun (Edible Fungi of China: a bimonthly journal)*, 7 (overall No. 29)(1): 28–31.

Zang, M. 1988b. An interesting edible mushroom: *Agaricus gennadii. Zhongguo Shiyongjun (Edible Fungi of China: a bimonthly journal)*, 7 (overall No. 32)(4): 3–4.

Zang, M. & Doi, Y. 1995. *Secotium jimalaicum* sp. nov. from Nepal – a folklore concerning the food of abominable snowman. *Acta Botanica Yunnanica*, 17(1): 30–32.

Zang, M. & Petersen, R. 1990. An endemic and edible fungus – *Endophyllus yunnanensis* from China. *Zhongguo Shiyongjun (Edible Fungi of China: a bimonthly journal)*, 9(3): 3–5.

Zang, M. & Pu, C. 1992. Confirmatory *Tuber indica* distributed in China. *Zhongguo Shiyongjun (Edible Fungi of China: a bimonthly journal)*, 11(3): 19.

Zang, M. & Yang, Z.L. 1991. *Agrocybe salicacola*, a delicious edible mushroom newly discovered from Yunnan. *Zhongguo Shiyongjun (Edible Fungi of China: a bimonthly journal)*, 10(6): 18.

Zeller, S.M. & Togashi, K. 1934. The American and Japanese matsu-takes. *Mycologia*, 26: 544–558.

Zerova, M.Y. & Rozhenko, G.L. 1988. *Atlas s'edobnykh i yadovitykh gribov [Atlas of edible and poisonous fungi – Ukraine]*. Kiev, Ukraine, Radyans'ka shkola. 40 pp.

Zerova, M.Y. & Wasser, S.P. 1972. *Edible and toxic mushrooms of the Carpathian forests*. Uzhorod, Karpaty Press. 128 pp.

Zhang, G. 1999. *Illustration for China popular edible mushroom*. Beijing, China Scientific Book Services. 110 pp.

Zhuang, Y. 1993. Characterization and textual criticism of Huai er (*Trametes robiniophila. Zhongguo Shiyongjun (Edible Fungi of China: a bimonthly journal)*, 13(6): 22–23.

Zhuang, Y. & Wang, Y.S. 1992. Applied research for second nutrition of *Gastrodia elata*. Quality of *Gastrodia elata* with trace element Zn. *Zhongguo Shiyongjun (Edible Fungi of China: a bimonthly journal)*, 11(6): 5–6.

ANNEX 1

Summary of the importance of wild edible fungi by region and country

GROUPS

Countries are arranged in six regions.

- Africa
- Asia
- Europe
- North and Central America [includes Caribbean region]
- Oceania
- South America

SOURCES OF INFORMATION

The country summaries highlight key information on wild edible fungi though details are often sparse, particularly on the broader social and economic contexts of use. Lists of "edible" species published in the mycological literature are of very limited use unless it is made clear which ones are actually eaten.

Two comprehensive reviews on wild fungi in Africa south of the Sahara have been particularly useful: Rammeloo and Walleyn (1993) for edible fungi and Walleyn and Rammeloo (1994) for poisonous and useful fungi. Key references are noted separately.

For many countries little or no published information on wild edible fungi was found. There are some clues to suggest that local use does occur but has yet to be described. No details of wild edible fungi use in Rwanda were found yet neighbouring Burundi has regular collecting, sale and consumption. Few details were found for Viet Nam and none for Myanmar yet there are cultural links to China, the country with the strongest tradition of wild edible fungi. Little information is available on Angola though it has large tracts of miombo woodland that are productive in neighbouring countries.

TRADE AND EXPORTS

Information is often incomplete and widely dispersed and trade data are missing for important exporting countries. Overall, the best information available is at www.fintrac.com but only covers 1993–97.

FUNGI THAT APPEAR ON STAMPS

A comprehensive description of all fungal species (mostly macrofungi) that have appeared on stamps since Romania produced the first examples in 1958 is available (McKenzie, 1997). Most of the 1 400 examples are edible species. Medicinal and poisonous varieties also appear. The list of species appearing on stamps is useful for countries where few other sources of information are available, for example the Democratic People's Republic of Korea. Small island nations exploit colourful species to increase revenue from stamp sales and the examples used are therefore a poor indication of local importance.

Africa

No information was found on wild edible fungi and other useful species for the following countries:

Cape Verde; Chad; Comoros; Djibouti; Equatorial Guinea; Eritrea; Gambia; Liberia; Mali; Mauritania; Niger; Sao Tome and Principe; Seychelles; St Helena; Sudan; Togo; Western Sahara

Two frequently cited reviews appear as: **R+W** (Rammeloo and Walleyn, 1993) and **W+ R** (Walleyn and Rammeloo, 1994).

For general information on NWFP in Africa see FAO (2001b). The only information found on fungi as emergency (famine) food concerned refugees from Mozambique who fled to Malawi in the 1980s (Wilson *et al.*, 1989).

Country	Use of Wild Edible Fungi
Algeria	Has exported matsutake in minor quantities to Japan, most likely *Tricholoma caligatum*. Desert truffles occur but few details are given (Alsheikh and Trappe, 1983). There are possibly exports to Spain (Borghi, 2002, personal communication: *Porcini and other commercial wild edible fungi in Italy*).
Angola	There is limited information that edible species are collected and used locally (FAO, 2001a). Isolated examples of wild edible species are given in **R+W.** Angola has miombo woodland similar to neighbouring countries where edible species are regularly collected and consumed. Further investigation is required.
Benin	Recent work reveals an extensive range of species that are consumed locally (De Kesel, Codjia and Yorou, 2002) and a long tradition of eating wild edible fungi. Few are openly sold.
Botswana	R+W lists a few species. Desert truffles are eaten and exported but harvests are very variable (Taylor *et al.*, 1995).
Burkina Faso	**R+W** lists a few species. A study of ectomycorrhizal fungi (Sanon, Ba and Dexheimer, 1997) confirms that edible species occur, though use as food is not discussed.
Burundi	Many different species occur and are collected and sold each year by rural people (Buyck, 1994b). There are distinct preferences for species among Africans and European expatriates.
Cameroon	Several reports and records have appeared and are summarized in **R+W**. No suggestion of major use of wild edible fungi but commonly collected and eaten.
Central African Republic	**R+W** list species from several sources. Forest dwellers appear to make the greatest use of wild fungi though this could reflect more detailed studies of these communities.
Congo [Republic of]	**R+W** has little information. A poorly studied country where wider use might be expected.
Congo, Democratic Republic of the (former Zaire)	Many publications and much research interest reveal widespread and significant use of wild edible species. Most reports concentrate on the Shaba region (e.g. Degreef, 1992). Information also in **R+W**.
Cote d'Ivoire	**R+W** list only a few records, but there are suggestions that use of wild edible fungi has been under-recorded and that several species are consumed and traded.
Egypt	Only one short account has been found (Zakhary *et al.*, 1983). No evidence to suggest that wild edible fungi are either abundant or routinely used.
Ethiopia	Only two short reports are known (Abate, 1999; Tuno, 2001). No evidence to suggest widespread use or importance of wild edible fungi.

COUNTRY	USE OF WILD EDIBLE FUNGI
GABON	**R+W** contains two records gleaned from earlier report which named 23 different types of WEF but using local names for most (Walker, 1931), suggesting common consumption.
GHANA	**R+W** contains few records. Information from the Forestry Research Institute of Ghana confirms that several species are collected and used (Obodai and Apetorgbor, 2001).
GUINEA	**W+R** has one record. Much wider use is expected and may have escaped detection because collection is essentially local and seasonal.
GUINEA-BISSAU	No information on wild edible found though a study of mycorrhizal fungi confirms the presence of edible varieties (Thoen and Ba, 1989).
KENYA	**R+W** and **W+R** contain several records but there is no evidence to support widespread collecting or trading.
LESOTHO	**R+W** has one record of a termite fungus. No other information available but note the presence of forest tree species (pines) associated with edible mycorrhizal fungi.
LIBYAN ARAB JAMAHIRIYA	Only one passing reference to desert truffles (Alsheikh and Trappe, 1983).
MADAGASCAR	**R+W** and **W+R** note several edible species though precise details of collection, consumption and sale are obscure (Bouriquet, 1970). No exports are known. More detailed studies are needed given the clear signs of major activities (Buyck, 2001).
MALAWI	A small country with a well-established tradition of using wild edible fungi. It has been well studied by comparison with similar countries **(R+W; W+R;** Morris, 1987; Boa *et al.*, 2000). See also www.malawifungi.org.
MAURITIUS	A few records exist (**R+W; W+R;** Peerally, 1979) but no details are available.
MOROCCO	Macrofungi are well-described and a range of edible species occur (Malencon and Bertault, 1975). Their significance to local people is not well known. It is a small-scale exporter of mushrooms (sic) to Japan, including a matsutake relative (*Tricholoma caligatum* – see Kytovuori, 1989).
MOZAMBIQUE	A country rich in edible species. These are routinely collected, consumed and sold internally but details are sketchy (Uaciquete, Dai and Motta, 1996; Boa *et al.*, 2000). Further study is required. There are also suggestions of B. edulis exports to Italy via companies based in South Africa (Borghi, 2002, personal communication: *Porcini and other commercial wild edible fungi in Italy*).
NAMIBIA	A few isolated records (**R+W** and **W+R**). No major use of wild edible species is indicated but there are regular exports of desert truffles (Taylor, 2002, personal communication: *Edible fungi eaten and traded in Botswana and Namibia*). Useful macrofungi occur in the Namib desert (Jacobson 1996).
NIGERIA	Brief lists of edible species are noted, mostly in connection with the Yoruba people (**R+W** and **W+R**). Several others reports exist (e.g. Oso, 1975) but they often repeat details published previously.
RWANDA	No records in **R+W** or **W+R** but information from Burundi (Buyck, 1994b) is relevant.
SENEGAL	Accounts of ectomycorrhizal species confirm that edible species are present (Thoen and Ba, 1989) but little is known about their use by local people (Ducousso, Ba and Thoen, 2002).
SIERRA LEONE	Only one passing reference (to Termitomyces) was found (Pegler and Vanhaecke, 1994). Mende women collect and sell edible fungi in Segbwema and presumably this occurs in other local markets (Down, 2002, personal communication: *Wild edible fungi Sierra Leone*). Further study is required.
SOMALIA	No information was found and there is no indication of widespread or regular use (**R+W**).
SOUTH AFRICA	Much mycological information but details on local non-European preferences and practices are only slowly being revealed (Shackleton *et al.*, 2002). See **R+W** and **W+R** for further discussions. Termitomyces collected and sold in KwaZulu (van der Westhuizen and Eicker, 1994). There are regular exports of *Boletus edulis* from pine plantations (Marais, 2002, personal communication: *Collecting* B. edulis *in South Africa*) which began in the 1970s (Pott, 2002, personal communication: *Export of* B. edulis *from South Africa*).

Country	Use of Wild Edible Fungi
Swaziland	Few details available about local use. Irregular exports of boletes in small quantities to Europe during the 1990s have occurred and appear to still take place (Borghi, 2002, personal communication: *Porcini and other commercial wild edible fungi in Italy*).
Tanzania [United Republic of]	**R+W** and **W+R** list many species. Good descriptions available of a wide range of edible fungi that are regularly collected, consumed and sold locally. Different species eaten in Miombo woodland and mountainous areas. An excellent and well illustrated guide to wild mushrooms has been published (Härkönen, Niemelä and Mwasumbi, 2003).
Tunisia	Only one short report on desert truffle was found (Alsheikh and Trappe, 1983). A minor and irregular exporter of "mushrooms", possibly to Spain (Borghi, 2002, personal communication: *Porcini and other commercial wild edible fungi in Italy*).
Uganda	**R+W** contains only a few records. A wider and stronger tradition is indicated (see Katende *et al.*, 1999). Information from Burundi is relevant (Buyck, 1994b).
Zambia	Widespread, common and significant use of wild edible species has been well described (e.g. (Pegler and Piearce, 1980; Piearce, 1981). **R+W** and W+R summarize records.
Zimbabwe	Wild edible fungi are commonly collected, sold and consumed. Boletus edulis is exported to Europe (Boa *et al.*, 2000). See also Ryvarden, Piearce and Masuka (1994) and **W+R**. Local traditions have been investigated in some detail only in the last 10 to 15 years and are less well described compared to Malawi and Zambia. Further attention is warranted.

Asia

No information was found on wild edible fungi and other useful species for the following countries or regions:

Azerbaijan; Bahrain; Brunei; Cambodia; Cyprus; Gaza Strip; Georgia; Kazakhstan; Maldives; Oman; Qatar; Syrian Arab Republic; Tajikistan; Timor-Leste; United Arab Emirates; Uzbekistan;West Bank; Yemen

The proximity of Azerbaijan and Georgia to countries with a known tradition of wild edible fungi (e.g. Armenia and Turkey) suggests a wider use of wild edible fungi than has been reported. Anecdotal information indicates that Kazakhstan has "little or no" tradition of wild edible fungi. The use of wild edible fungi in Tajikistan and Uzbekistan is expected but has yet to be confirmed. So too for Cambodia: there is a tradition among tribal people in the region of using wild edible fungi (Hosaka, 2002, personal communication: *Laos edible fungi*) See Plates 8 and 9.

Country	Use of Wild Edible Fungi
Afghanistan	A few wild edible species are described (Batra, 1983). Morels are exported (Sabra and Walter, 2001).
Armenia	A range of available edible species are collected, consumed and traded locally. Exports have not been reported (Nanaguylan, 2002, personal communication: *Edible fungi in Armenia*).
Bangladesh	Small-scale use by Chakma people in Hill Tracts has been noted (Siddiqi, 1998).
Bhutan	A small-scale exporter of matsutake to Japan but important to the local economy. Wild edible species are regularly sold in markets though species and amounts are not known (Namgyel, 2000).
China	The leading producer, user and exporter of wild edible fungi in the world with a long and notable tradition of using medicinal species. There are significant exports of matsutake to Japan though harvesting practices are causing concern for sustainable production in some areas (Winkler, 2000). Truffles and *Boletus edulis* exported more recently in significant quantities to Europe (Borghi, 2002, personal communication: *Porcini and other commercial wild edible fungi in Italy*). General lists of species in regular use have been published outside China (e.g. Hall *et al.*, 1998a) but should be consulted together with an expanding Chinese Literature. See Mao and Jiang, 1992 for Tibet Autonomous Region; Ying *et al.*, 1987; Ying *et al.*, 1988. *Zhongguo Shiyongjun* [Edible Fungi of China] regularly publishes information but in Chinese. Few accounts of fungi sold in markets have been published (Chamberlain, 1996) though this is a widespread and important activity. For medicinal species generally see Hobbs (1995). The best guide and source of information on field mycology and species of WEF is Mao, 2000).
Hong Kong Special Administrative Region, China	Chang and Mao (1995) is a comprehensive account of macrofungi and their useful characteristics (in Chinese). This has a wider relevance to China.

India	Lists of edible species from the extensive mycological records are difficult to interpret and social and economic aspects are poorly studied. For general information see Purkayastha and Chandra, 1985. Studies of local use include: Harsh, Rai and Ayachi, 1993; Harsh, Rai and Soni, 1999; Adhikary *et al.*, 1999. Morels are collected for export in Himalayan regions (FAO, 1993b) and are of economic importance. Further studies are needed, particularly in hill areas where tribal people live, e.g. Tripura and Mizoram.

Country	Use of Wild Edible Fungi
Indonesia	Very little information has been published though there is clear evidence of widespread use and market selling (Burkhill, 1935; Heyne, 1927; Rifai, 1989). There is much interest in cultivating fungi (e.g. Gunawan, 2000) and these are widely available. The extensive literature on NWFP has few details of wild edible fungi though local sources in Kalimantan (Leluyani, 2002, personal communication: *Edible fungi of Kalimantan*) listed over ten different types regularly collected and consumed in forest areas, mostiy saprobic species. Canned *Scleroderma* spp. are sold (Ducousso, Ba and Thoen, 2002). Published records of agarics and boletes are available at www.mycena.sfsu.edu and include several common edible species.
Iran	Truffles occur but their significance to local people is not known (Saremi, Ammarellou and Mohammadi, 2002). Other edible and medicinal species have been recorded (see Niemelä and Uotila, 1977; Isiloglu and Watling, 1992) in the mycological literature.
Iraq	Only one passing reference [to desert truffles] is known (Al-Naama, Ewaze and Nema, 1988).
Israel	The recent arrival of many Russians has introduced a strongly mycophilic influence (Wasser, 1995), though there is still little available information on how collection and consumption of wild edible fungi has changed. Previously there was only limited interest in a few key species.
Japan	It has a notable and significant tradition of collecting, consuming and selling wild useful fungi (e.g. Kawagoe, 1924; Stamets, 2000). There is an extensive literature on macrofungi (e.g. Imazeki *et al.*, 1988) and research on wild edible species, particularly *matsutake*. Japan is a major importer of *matsutake* and related species from around the world.
Jordan	Several species are consumed locally (Cavalcaselle, 1997; Sabra and Walter, 2001).
Korea [Democratic People's Republic of]	There is undoubtedly a strong local tradition of collecting and consuming wild edible fungi but information is scarce. There are significant exports of matsutake to Japan (www.fintrac.com).
Korea [Republic of]	It has a strong local tradition of using wild edible fungi and is a major exporter of matsutake to Japan. For further information, see Kim and Kim (1990).
Kuwait	Only one account with a passing reference to desert truffles is known (Alsheikh and Trappe, 1983).
Kyrgyzstan	A comprehensive list of edible species has been published (El'chibaev, 1964) which suggests widespread if not necessarily significant use of wild species.
Lao People's Democratic Republic	A list of edible species with photos is available at http://giechgroup.hp.infoseek.co.jp/kinoko/eng.html. NWFP studies include references to wild edible fungi (Rijsoort and Pikun, 2000). Local use is widespread (Hosaka, 2002, personal communication: *Laos edible fungi*) but poorly described. Further studies are needed to reveal more details about the use of wild edible fungi by hill people generally in the region.
Lebanon	Several species are locally collected though apparently use is small scale and may not be widespread (Sabra and Walter, 2001).
Malaysia	Termite fungi are regularly collected and sold (Pegler and Vanhaecke, 1994). Mycological reports from Sarawak (Chin, 1988; Chin, 1998) hint at regular use of wild edible species, confirmed by anecdotal accounts (Jones, 2002, personal communication: *Wild edible fungi use in Sarawak*).
Mongolia	No information was found but similar traditions to neighbouring countries (e.g. China) are expected.
Myanmar	Termite fungi are recorded in the mycological literature (Pegler and Vanhaecke, 1994) and are undoubtedly eaten, but no other details have been found. However, similar patterns of use are expected in the hill regions based on traditions in neighbouring countries.

Nepal	Widespread collection, sale and consumption occur (e.g. Adhikari and Adhikari, 1996), with most activity in the hill regions.
Pakistan	Only limited information was found. Morels are collected and exported (FAO, 1993b). Mycological reports do not describe local practices or preferences for species (Batra, 1983; Syed-Riaz and Mahmood-Khan, 1999).

Country	**Use of Wild Edible Fungi**
Philippines	A comprehensive mycological paper (Mendoza, 1938) lists over 50 species, many with local names and suggesting widespread use. This information is not included in the annexes. Forest dwellers in Palawan also eat wild edible fungi (Novellino, 1999).
Saudi Arabia	Limited information on desert truffles (Tirmania) only was found (Bokhary and Parvez, 1993).
Singapore	A significant importer and user of edible fungi though mostly, it is suspected, of the cultivated species (Jones and Lim, 1990). A strong cultural influence from the Chinese tradition is expected.
Sri Lanka	Local collections occur but limited information was found (Gunatilleke, Gunatilleke and Abeygunawardena, 1993). Termite fungi occur and are presumably eaten (Pegler and Vanhaecke, 1994).
Taiwan Province of China	Similar tradition to mainland China though information not actively gathered. Long tradition of mycological research on the higher fungi (see Chen, 1987).
Thailand	There is a notable tradition of collection, selling and consumption but only one detailed report was found (Jones, Whalley and Hywel-Jones, 1994).
Turkey	There is a strong but perhaps still relatively small export industry to Europe, based predominantly on the collection of wild edible fungi (Gurer, 2002, personal communication: *Unpublished trade data on wild edible fungi in Turkey*). Mycological reports suggest widespread use and significance (e.g. Afyon, 1997; Kasik and Ozturk, 1995). See also www.ogm.gov.tr/ and Sabra and Walter (2001).
Turkmenistan	Has exported "mushrooms" to Germany, most probably wild edible species (www.fintrac.com).
Viet Nam	There are clear indications of widespread local use and collecting in the upland areas (Chamberlain, 2002, personal communication: *Wild edible fungi in Viet Nam*) but this is poorly documented. NWFP investigations frequently mention wild edible fungi (e.g. Rijsoort and Pikun, 2000). Paddy straw (*Volvariella* spp.) occurs naturally in lowland areas and is also cultivated. Other cultivated species such as *shiitake* and ear fungi (*Auricularia* spp.) are sold fresh and dried in markets in Ho Chi Minh city.

Europe

The macrofungi of Europe, as defined by the present boundaries of the European Union and contiguous countries, are well known and described. Finland has the most comprehensive literature on collection and use of edible fungi and has paid particular attention to their importance for people.

Information on edible fungi from Liechtenstein, Malta and Iceland was not found.

Countries fall in to two broad groups: first, nations with weak economies, usually with a significant local tradition of using wild edible fungi and some which also export; second, wealthier countries that import but may not have a strong tradition of collecting. Romania is an example of the first group and the Netherlands an example of the second. (The Netherlands is the largest global exporter of button mushrooms – *Agaricus bisporus* – and third exporter after China and the United States of all cultivated species.)

The easing of economic and political barriers in the early 1990s has stimulated exports from former Soviet countries, Balkan states and Yugoslavia specifically (Perini, 1998). Within the richer countries of Europe collecting wild edible fungi is mostly for small-scale personal use and is of minor economic importance to the collectors, though there is a growing individual interest in collecting truffles and *porcini* in Italy (Zambonelli, 2002, personal communication: *Truffles, and collecting porcini in Italy*). See Plates 3 and 4.

For accounts of wild edible fungi collected from boreal and cold temperate forests see Lund, Pajari and Korhonen (1998).

Country	Use of Wild Edible Fungi
Albania	It has exported limited quantities of edible fungi to Italy, probably *Boletus edulis* (Borghi, 2002, personal communication: *Porcini and other commercial wild edible fungi in Italy*) and a few other types, but there is no regular trade.
Belarus	Wild edible species are described briefly (Malyi, 1987) but without details of local practices. Also exports wild species in small quantities to Italy (Borghi, 2002, personal communication: *Porcini and other commercial wild edible fungi in Italy*) and other unspecified countries (Ollikainen, 1998).
Bosnia and Herzegovina	Exports "mushrooms", including *Boletus edulis* to Italy (Borghi, 2002, personal communication: *Porcini and other commercial wild edible fungi in Italy*). No other information or reports have been seen.
Bulgaria	Major exporter of "wild mushrooms". Edible and poisonous species have been described in the mycological literature (Iordanov, Vanev and Fakirova, 1978) though local traditions are not well known.
Czech Republic	A minor exporter to neighbouring Germany, assumed to be mostly from the wild. Local collecting and consumption was regulated some time ago (Pilát, 1951) and appear to be mostly for internal consumption (Sisak, 1998).
Croatia	Exporter but activities disrupted by civil strife. Exact details are unclear but see comments for Serbia and Montenegro.
Estonia	Known to have a strong tradition of local use and research on wild edible fungi (Kalamees and Silver, 1988). Production data indicate it is a minor exporter (Paal and Saastamoinen, 1998), at least from 1993 to 1997 (www.fintract.org).

COUNTRY	USE OF WILD EDIBLE FUNGI
FINLAND	Traditions vary from the mycophilic east, influenced by its proximity to the Russian Federation, to the less enthusiastic west, taking its influences from Sweden (Härkönen, 1998). There has been official encouragement to collect edible fungi since the Second World War and discussions and research on inventory and long-term yield studies (Rautavaara, 1947; Koistinen, 1978); access to lands (Saastamoinen, 1999); local mushroom advisors (Mildh, 1978; Härkönen, 1988).
GREECE	Commonly collected and used in rural areas from forests (Diamandis, 1997). Few are sold in farmers' markets though there have been increases in commercial picking which are causing concern (Diamandis, 2002). Have been eaten since ancient times (Hettula, 1989).
HUNGARY	Exports and has a local tradition of collection and consumption, but few published details are available apart from lists of species (Grunert and Grunert, 1995).
ITALY	Extensive imports of *Boletus edulis* (*porcini*) from a wide range of countries, extending to China (over 60% of imports according to Borghi, 2002, personal communication: *Porcini and other commercial wild edible fungi in Italy*) and southern Africa. See (Hall *et al.*, 1998b) for general information on *porcini*. Recently, an inferior *Tuber* from China has been imported (Hall, Zambonelli and Primavera, 1998a; Zang and Pu, 1992). See Buller (1914) for historical perspective. In the past the collection of wild edible fungi was important to the livelihoods of many people in the northern regions. While there is still a strong interest in collecting and eating, particularly *porcini* and truffles, their economic importance to local people has declined. Still, there is a strong commercial interest in both groups of fungi with demand outstripping local supply (Borghi, 2002, personal communication: *Porcini and other commercial wild edible fungi in Italy*). Italy has an impressive mycological tradition but there is a paucity of information on local traditions and uses of WEF by people.
LATVIA	Relatively minor exporter at least from 1993 to 1997 (www.fintrac.com). It has a similar local tradition of use compared to Estonia and Lithuania (Vilkriste, 1998). For selected list of edible species see Urbonas, Kalamees and Lukin (1974).
LITHUANIA	Major exporter to Germany over the period 1993 to 1997 but in variable quantities (www.fintrac.com). Around 190 edible species are listed by Butkus *et al.* (1987). Further information available in Rutkauskas (1998).
MACEDONIA [THE FORMER YUGOSLAV REPUBLIC OF]	Regular exporter, including *Boletus edulis* to Italy (Borghi, 2002, personal communication: *Porcini and other commercial wild edible fungi in Italy*) and with a suggested strong tradition of local use (Bauer-Petrovska *et al.*, 2001).
MOLDOVA	Minor exports of *Boletus edulis* to Italy (Borghi, 2002, personal communication: *Porcini and other commercial wild edible fungi in Italy*). Likely to have a similar tradition of collecting and use to the Russian Federation.
POLAND	Europe's leading exporter of "mushrooms" and a major source of revenue. It is said to be a the pioneer in protecting wild edible fungi with legislation introduced in 1983 (Lawrynowicz, 1997). Also has a strong local tradition in the poorer regions (Snowarski, 2002, personal communication: *Wild edible fungi in Poland*). For general information see www. grzyby.pl and Kalinowski (1998).
ROMANIA	Major exporter of wild edible fungi (Pop, 1997), with *Boletus edulis* sent to Italy on a regular basis (Borghi, 2002, personal communication: *Porcini and other commercial wild edible fungi in Italy*).
RUSSIAN FEDERATION	A strong and lengthy tradition of collecting and consuming wild edible fungi exists (Wasson and Wasson, 1957). Precise details of current use are difficult to find though there is an impressive mycological literature and history of research on species (e.g. Dudka and Wasser, 1987; Vasil'eva, 1978; Wasser, 1990). It is the second most important country or region for wild edible fungi after China in terms of amounts collected but trails in value of exports – though these have occurred for many years (Paal, 1998). There is a certain fearlessness in picking fungi as indicated by regular poisoning and even deaths (Chibisov and Demidova, 1998; Evans, 1996). Concern has been expressed about rampant exports in "hundreds of tons", with St Petersburg a "much exploited region" (Kovalenko, 1997).
SERBIA AND MONTENEGRO [FORMER YUGOSLAVIA]	Exports of *Boletus edulis* to Italy began in the 1970s (Borghi, 2002, personal communication: *Porcini and other commercial wild edible fungi in Italy*) and regularly ever since. Exports increased significantly in the 1990s, of *B. edulis* and other species, with significant rises in the numbers of people earning a living from commercial activities (Ivancevic, 1997). In sharp contrast, there are weak local traditions of use (Žaklina, 1998).

Country	Use of Wild Edible Fungi
Slovakia	Unconfirmed reports of widespread collecting are similar to traditions in neighbouring countries, for example Poland.
Slovenia	Moderate amounts are exported, including *Boletus edulis* to Italy. It has a notable though not necessarily strong local tradition (www.matkurja.com).
Spain (and Andorra)	Sharply differing traditions of local use with the strongest existing among the mushroom-loving Catalans and also Basque people. Their interests drive much of the internal trade in WEF. There is an important trade in *Lactarius deliciosus* (níscalos) from northwest Spain (Castilla y Leon) to Catalonia while truffles are of increasing importance to local people in the Pyrenees (de Román, 2002, personal communication: *Trade in níscalos from North Spain to Catalonia and truffle production*). For a comprehensive account of wild edible fungi see Martínez, Oria de Rueda and Martínez (1997). Spanish traders visit Portugal for commercial activities while French collectors cross over to Spain for truffles. See also (Wasson and Wasson (1957) for historical information on local traditions.
Ukraine	Possesses significant resources that are highly valued by local people (Zerova and Wasser, 1972; Zhang, 1999). There has been much concern expressed about contamination by radioactive materials following the Chernobyl accident but this is overshadowed by the dramatic rise in deaths from eating poisonous species (Vachuska and Vachuska, 2000), events linked to a weak economy and a desperate search for food (Almond, 2002).

Collections in the following countries are essentially for occasional individual use. General comments concern exports and imports, depending on available information.

Country	Use of Wild Edible Fungi
Belgium and Luxembourg	Exports species but details vague. Scientists have made major contributions to African Mycology (Rammeloo, 1994).
Denmark	Small-scale and infrequent local collections only (Plum, 1998).
France	Major importer and exporter (sometimes from third party countries e.g. Portugal, Spain). At one time exported large quantities of truffles to Italy (Ainsworth, 1976). There is a strong tradition of collecting and eating WEF in the south (e.g. Gascony, Provence) but published information on local traditions has not been found
Germany	Major importer of wild edible fungi, e.g. chanterelles.
Ireland	Major exporter but mostly (only?) cultivated species to the United Kingdom (www.fintrac.com).
Netherlands	Europe's leading exporter of mushrooms, mostly cultivated species.
Norway	Common edible species such as chanterelles and boletes are collected for personal use.
Portugal	Local traditions are weak (Martins *et al*., 2002) and this has been exploited by traders from Spain and France who have created a "flourishing and uncontrolled commercial" business (Baptista-Ferreira, 1997): hundreds of tonnes of *Boletus edulis* and related species are exported.
Sweden	Chanterelles and other common edible species are sold but there is no strong tradition of collecting. There is an increased interest in cultivating truffles.
Switzerland	There is fierce competition by collectors for local resources (see Egli, Ayer and Chatelain, 1990). Some information on imports of wild edible fungi is presented in Wills and Lipsey (1999).
United Kingdom	Major importer of mushrooms, particularly from Ireland (see www.fintrac.com). Small-scale commercialization of wild edible fungi has begun and there is a useful study of collectors and the developing trade (Dyke and Newton, 1999). Concerns about overpicking and damage caused by collectors has led to the introduction of local regulations at several sites in southern England (e.g. New Forest, Epping Forest).

North and Central America

See Plate 7. No information was found on wild edible fungi and other useful species for the following countries:

Antigua and Barbuda; Antilles, Netherlands; Bahamas; Barbados; Belize; Bermuda; British Virgin Islands; Cayman Islands; Dominica; Dominican Republic; Grenada; Guadeloupe; Martinique; Monserrat; Nicaragua; Panama; Puerto Rico; Saint Kitts and Nevis; Saint Lucia; Saint Pierre and Miquelon; Saint Vincent and the Grenadines; Trinidad and Tobago; United States Virgin Islands

Country	Use of Wild Edible Fungi
Canada	Exports to Japan and to Europe. Several publications described the expansion in collection and trade of wild edible fungi, principally from British Colombia (the "Pacific northwest") (see Redhead, 1997; Tedder, Mitchell and Farran, 2000). Some United States publications include Canada in their discussions (Pilz and Molina, 2002). First nation people have collected and used for many years (Marles *et al.*, 2000).
Costa Rica	Studies on the diversity of macrofungi are well advanced, though without any clear emphasis on edible species (Mata-Hidalgo, 1999). Lists of edible and poisonous species (Saenz, Lizano and Nassar, 1983) confirm weak local traditions.
Cuba	There is little or no apparent tradition of using wild edible fungi (Minter, 2002, personal communication: *Edible fungi in Chile, Cuba and Argentina*).
El Salvador	Exports to Germany but irregular and small scale. Intensive agriculture and deforestation suggests few collections are made though note strong tradition in nearby Guatemala.
Guatemala	Strong tradition in the Western Highlands (Flores, 2002, personal communication: *Guatemala edible fungi*; Flores, Bran and Honrubia, 2002; de Leon, 2002). An account of poisoning (Logemann *et al.*, 1987) points to the wider significance of wild edible fungi though again mainly in the western highlands. Local edible species have been documented (Sommerkamp and Guzmán, 1990) and historical accounts of use exist (e.g. Lowy, 1971).
Haiti	Haitian expatriates regularly buy *djon djon*, a *Psathyrella* species (Nieves-Rivera, 2001), which is cultivated only in Haiti (Yetter, 2002, personal communication: *Edible fungi from Haiti; for sale in Brooklyn; link to eating* Psathyrella *in Africa*) and exported around the world. Local details of production are sketchy. A few other wild edible fungi are collected and some information is available in Alphonse, 1981, but this reveals few details.
Honduras	Extensive areas of natural pine forest are associated with good wild edible fungi. There is a tradition in the west, close to the border with Guatemala, where around three or four species are sold in local markets (House, 2002, personal communication: *Wild edible fungi in Honduras*).
Jamaica	Minor and irregular exports of "mushrooms" to Germany (www.fintrac.com) but details are sketchy. There is no obvious tradition of wild edible fungi in the Caribbean with the major known exception of Haiti.

Country	Use of Wild Edible Fungi
Mexico	One of the most important countries for use and significance of collections to local people. It is unusual in the extent to which this has been described by local scientists (see Villarreal and Perez-Moreno (1989) for a good summary). For good online access to key information see SEMARNAT (2002). Small-scale exports of selected species. Wild fungi also play a strong cultural role (Riedlinger, 1990). There is a vigorous body of researchers working on wild edible fungi and regular publications that are now turning their attention to key social and economic issues.
United States	Major exporter to Japan of *matsutake* but also a notable importer from a wide range of places. Has a rich literature and tradition in mycological sciences and is the academic "home" of ethnomycology (see Schultes 1940; Riedlinger 1990). The tradition of local use and collections is much less than that suggested by the vast scientific canon. That which does exist owes much to the cultural background of immigrants from Europe and Japan (less is known about the influence of Chinese immigrants; see also notes above on Haiti). However, there are also significant accounts by native Americans (e.g. Keewaydinoquay, 1998). Most recent interest has centred on the export-driven collections and subsequent huge expansion of commercial activities and trade centering around the Pacific northwest. This trade has been stimulated by a decline in forestry jobs and the demand for *matsutake* in Japan. There is an extensive literature on this topic (see Pilz and Molina, 2002 for a comprehensive review).

Oceania

No information was found on wild edible fungi and other useful species for these countries:

Cook Islands; French Polynesia; Guam; Kiribati; Marshall Islands; Micronesia; Nauru; New Caledonia; Niue; Northern Mariana Islands; Palau; Samoa; American Samoa, Solomon Islands; Tonga; Vanuatu

COUNTRY	USE OF WILD EDIBLE FUNGI
AUSTRALIA	There is useful account of aboriginal use (Kalotas, 1997).
FIJI	A brief account (Markham, 1998) describes a weak tradition of collecting from the wild.
NEW ZEALAND	Most notable for the successful research and development efforts in cultivating *Tuber* spp. (see Hall *et al.* (1998a) for general information). Once exported relatively large amounts of *Auricularia* to China (Colenso, 1884–85).
PAPUA NEW GUINEA	An informative ethnomycological study of one group of highland people hints at a more widespread importance (Sillitoe, 1995). An account of wild edible fungi used by the Gadsup people also lists many species used locally (Shaw, 1984), including "Amanitas and Russulas", but the original sources of this information (Heim, 1964) has not been seen.

South America

There are few comprehensive accounts of wild edible fungi for the region but note two papers that present useful information: first, Paraná in Brazil (Meijer, 2001) and, second, the Mercosur region comprising Argentina, Chile and Uruguay (Deschamps, 2002). See Plate 7.

No information was found for these countries:

French Guiana; Guyana; Paraguay; Suriname; Venezuela

Country	Use of Wild Edible Fungi
Argentina	Morels are collected and sold locally, and there are commercial collections of *Suillus luteus* near Bariloche (Gamundí, 2002, personal communication: *Edible fungi collected in Argentina*). *Cyttaria* species are eaten in the south (Minter, Cannon and Peredo, 1987). A recent overview of wild edible fungi in the Mercosur regions has been published (Deschamps, 2002).
Bolivia	No information found on local use. An Indian lady was selling *Leucoagaricus hortensis* in Cochabamba market in March 2001 and suggests that some collection occurs (personal observation). The vendor was the only person offering wild fungi for sale (and in quantities of less than a kilogram).
Brazil	A country with a rich mycological tradition in science but weak tradition in use of wild edible fungi. Ethnomycological studies in Amazonia (Prance, 1984) reveals small-scale but important use that hints at wider collections for other forest dwellers in Colombia, Bolivia, Peru and Venezuela. Despite significant Italian migration to Rio Grande do Sul there is no reported collections, even though pines are widely planted (Schifino-Wittmann, 2002, personal communication: *Eating fungi in south Brazil*). The influence of a large ethnic Japanese population is also curiously muted though *Agaricus blazei*, a medicinal species, was apparently first discovered by someone of Japanese descent. The fungus is exported to Japan. The small-scale use of wild edible fungi among Europeans is commented on by Meijer (2001).
Chile	*Suillus luteus* is exported from forest plantations (see FAO, 1998a). There is a local Indian tradition [Mapuche] of eating *Cyttaria*, a curious golf-ball like fungus parasitic on *Nothofagus* (Minter, Cannon and Peredo,1987). A comprehensive list of fungi eaten locally is available (FAO, 1998b) and earlier information provides details of harvesting operations in Region VII (FAO, 1993a).
Colombia	A recent guide to macrofungi (Franco-Molano, Aldana-Gomez and Halling, 2000) includes edible species but has no information on local practices in the Andean region.
Ecuador	Irregular and small-scale exporter of pine boletes, principally if not wholly to the United States (Rojas and Mansur, 1995). *Suillus luteus* is the principal species involved (Hedger, 1986).
Peru	A preliminary list of wild edible fungi does not have details of local practices (Remotti and Colan, 1990). An ethnoscientific study suggests widespread collections by rural people (Franquemont *et al.*, 1990).
Uruguay	A recent overview of wild edible fungi has been published (Deschamps, 2002). This lists several species that are traded (see Annex 2).

ANNEX 2

Country records of wild useful fungi (edible, medicinal and other uses)

This list includes over 2 800 records from 85 countries and was prepared from a preliminary database record of published information. Information from the Republic of Korea, Japan and Taiwan Province of China is not included and records from European countries are limited (Box 2). The mycological literature is extensive in many developed countries but often there is no clear indication of which species are eaten as "food". The United States and Canada records are from the Pacific northwest region or from reports on first nation people (aboriginals). Australia records are for aboriginal use only.

Unnamed species are excluded unless there is no other named species for that genus from a particular country. Thus *Agaricus* sp. is not included if *Agaricus campestris* has been recorded.

Only uses of practical or economic importance have been included; ceremonial or religious uses are omitted.

Square brackets e.g. [edible], indicate uncertainty about the use in the source of the information.

Taxonomists use various ways to qualify a species names: cf. and aff. indicate that the specimen examined was close to the species name given (e.g. *Amanita* aff. *rubescens*) but they are not 100 percent certain. The letters s.l. mean *sensu lato* or "in the broad sense".

A complete list of all species and countries can be searched at www.wildusefulfungi.org. This contains all details about recorded uses and properties and includes Japan and Russia (Sergeeva, 2000) and a comprehensive list of wild edible fungi from China (Mao, 2000). This searchable database currently holds over 6 000 records from 108 countries and provides valid names of species.

AFGHANISTAN **1.** Batra, 1983; **2.** Sabra and Walter, 2001	
Morchella	edible (2)
Podaxis pistillaris	edible (1)

ALGERIA **1.** Alsheikh and Trappe, 1983; **2.** Kytovuori, 1989	
Tirmania nivea	edible (1)
Tirmania pinoyi	edible (1)
Tricholoma nauseosum	edible (2)

ANGOLA Rammeloo and Walleyn, 1993	
Macrolepiota procera	edible
Termitomyces sp.	edible

ARGENTINA **1.** Deschamps, 2002; **2.** Gamundí and Horak, 1995	
Cyttaria hariotii	food (2)
Morchella elata	food (1)
Morchella intermedia	food (1)
Phlebopus bruchii	food (1)
Suillus luteus	food (2)

ARMENIA Nanaguylan, 2002, personal communication	
Agaricus bisporus	food
Agaricus campestris	food
Agaricus silvaticus	food
Armillaria mellea	food
Calocybe gambosa	food
Cantharellus cibarius	food
Lactarius deliciosus	food

Lepista nuda	food
Lepista personata	food
Macrolepiota excoriata	food
Macrolepiota procera	food
Pleurotus eryngii	food
Pleurotus ostreatus	food
Suillus granulatus	food
Suillus luteus	food

AUSTRALIA Kalotas, 1997	
Battarrea stevenii	not known
Boletus sp.	edible
Choiromyces aboriginum	food
Cyttaria gunnii	food
Fistulina hepatica	food
Montagnites candollei	not known
Mycoclelandia bulundari	food, medicinal
Phellinus rimosus	medicinal
Phellorinia herculeana	other – dye
Phellorinia strobilina	not known
Pisolithus tinctorius	food, medicinal
Podaxis pistillaris	other – cosmetic
Polyporus eucalyptorum	food, tinder
Polyporus mylittae	food
Pycnoporus sanguineus	medicinal
Secotium sp.	medicinal

BELARUS Malyi, 1987	
Armillaria mellea	edible
Boletus edulis	edible
Cantharellus cibarius	edible
Gyromitra esculenta	edible
Lactarius deliciosus	edible
Lactarius necator	edible
Lactarius torminosus	edible
Leccinum aurantiacum	edible
Leccinum scabrum	edible
Morchella esculenta	edible
Suillus luteus	edible
Tricholoma flavovirens	edible
Tricholoma portentosum	edible
Xerocomus subtomentosus	edible

BENIN **1.** Antonin and Fraiture, 1998; **2.** De Groote, 2002; **3.** De Kesel, 2002, personal communication; **4.** De Kesel Codjia and Yorou, 2002; **5.** Walleyn and Rammeloo, 1994; **6.** Yorou and De Kesel, 2002; **7.** Yorou et *al.*, 2002	
Agaricus bisporus	food (6)
Agaricus bulbillosus	food (4)
Agaricus goossensiae	food (4)
Agaricus volvatulus	food (4)
Agrocybe howeana	food (3)
Amanita aff. rubescens	food (4)
Amanita craseoderma	food (4)
Amanita crassiconus	food (4)
Amanita loosii	food (6)
Amanita masasiensis	food (4)
Amanita strobilaceovolvata	food (4)
Amanita subviscosa	food (4)
Amanita xanthogala	food (4)
Auricularia cornea	food (4)
Boletus pseudoloosii	food (4)
Boletus sp.	food (3)
Calvatia subtomentosa	food (3)
Cantharellus congolensis	food (4)
Cantharellus floridulus	food (4)
Cantharellus platyphyllus	food (4)
Chlorophyllum cf. molybdites	food (4)
Clitocybe s.l. sp.	food (3)
Clitocybula sp.	food (3)
Craterellus beninensis	food (4)
Craterellus cornucopioides	food (3)
Daldinia concentrica	medicinal (3)
Gerronema sp.	food (3)
Gymnopus luxurians	food (6)
Hebeloma termitaria	food (4)
Inocybe gbadjii	food (3)
Inocybe squamata	food (6)
Lactarius baliophaeus	food (4)
Lactarius densifolius	food (4)
Lactarius edulis	food (4)
Lactarius flammans	food (4)
Lactarius gymnocarpoides	food (4)
Lactarius latifolius	food (3)
Lactarius luteopus	food (4)
Lactarius pseudogymnocarpus	food (6)
Lactarius pumilus	food (3)
Lactarius saponaceous	food (4)
Lactarius species 1	food (3)
Lactarius species 7	food (3)
Lactarius tenellus	food (4)
Lactarius volemoides	food (3)
Lentinus sp.	food (3)
Lentinus tuber-regium	food (4)
Lentinus velutinus	food (3)
Lentinus squarrosulus	food (4)
Lepista dinahouna	food (3)
Lepista sp.	food (3)
Leucoagaricus bresadolae	food (4)
Leucoagaricus sp. nov.?	food (2)
Leucoagaricus sp.	food (3)
Lycoperdon sp.	food (3)
Macrocybe lobayensis	food (4)
Marasmius becolacongoli	food (3)
Marasmius heinemannianus	edible (1)
Marasmius heinemannianus	food (4)
Marasmius spp.	food (3)
Nothopanus hygrophanus	food (3)
Octaviania ivoryana	food (4)
Phlebopus sudanicus	food (4)
Pleurotus cystidiosus	food (4)
Pleurotus djamor	food (3)
Pleurotus sp.	food (3)
Polyporus sp.	medicinal (5)
Psathyrella sp.	food (2)
Psathyrella tuberculata	food (4)
Rubinoboletus roseo-albus	food (3)

Russula aff. virescens	food (3)
Russula cellulata var. *nigra*	food (4)
Russula cellulata	food (4)
Russula compressa	food (6)
Russula congoana	food (4)
Russula grisea	food 7
Russula meleagris	food (4)
Russula oleifera	food (4)
Russula pseudopurpurea	food (6)
Russula testacea	food (6)
Schizophyllum commune	food (4)
Termitomyces aurantiacus	food (4)
Termitomyces clypeatus	food (4)
Termitomyces fulginosus	food (4)
Termitomyces letestui	food (4)
Termitomyces medius	food (4)
Termitomyces microcarpus	food (4)
Termitomyces robustus	food (4)
Termitomyces schimperi	food (4)
Termitomyces striatus	food (4)
Tylopilus sp.	food (3)
Volvariella earlei	food (4)
Volvariella volvacea	food (4)

BHUTAN
Namgyel, 2000

Albatrellus sp.	[edible]
Calocera viscosa	[edible]
Cantharellus cibarius	edible
Coprinus sp.	[edible]
Gomphus floccosus	edible
Hygrophorus russula	[edible]
Lactarius hatsudake	[edible]
Lactarius piperatus	edible
Lycoperdon pyriforme	edible
Lyophyllum fumosum	[edible]
Ramaria sp.	[edible]
Suillus pictus	[edible]
Tricholoma matsutake	food

BOLIVIA
Boa, 2001, personal communication

Leucoagaricus hortensis	food

BOTSWANA
1. Rammeloo and Walleyn, 1993; 2. Taylor *et al.*, 1995

Morchella conica	edible (1)
Terfezia boudieri	edible (1)
Terfezia pfeilii	food (2)

BRAZIL
1. Prance, 1984; 2. www.agaricus.net

Agaricus blazei	medicinal (2)
Auricularia fuscosuccinea	food (1)
Collybia pseudocalopus	food (1)
Collybia subpruinosa	food (1)
Favolus brasiliensis	food (1)
Favolus brunneolus	food (1)
Favolus striatulus	food (1)
Favolus tesselatus	food (1)
Gloeoporus conchoides	food (1)
Gymnopilus earlei	food (1)
Gymnopilus hispidellus	food (1)
Hydnopolyporus palmatus	food (1)
Lactocollybia aequatorialis	food (1)
Lentinus crinitus	food (1)
Lentinus glabratus	food (1)
Lentinus strigosus	food (1)
Lentinus velutinus	food (1)
Leucocoprinus cheimonoceps	food (1)
Neoclitocybe byssiseda	food (1)
Pholiota bicolor	food (1)
Pleurotus concavus	food (1)
Polyporus aquosus	food (1)
Polyporus indigenus	food (1)
Polyporus sapurema	food (1)
Polyporus stipitarius	food (1)
Polyporus tricholoma	food (1)
Pycnoporus sanguineus	food (1)
Trametes cubensis	food (1)
Trametes ochracea	food (1)
Trichaptum trichomallum	food (1)

BULGARIA
Iordanov, Vanev and Fakirova, 1978

Agaricus arvensis	[edible]
Agaricus aurantius	not known
Agaricus bulbosus	not known
Agaricus campestris	[edible]
Agaricus comptulus	not known
Agaricus maculatus	not known
Agaricus pseudoaurantiacus	not known
Agaricus silvaticus	[edible]
Albatrellus confluens	[edible]
Albatrellus ovinus	[edible]
Amanita argentea	[edible]
Amanita caesarea	[edible]
Amanita fulva	[edible]
Amanita pustulata	not known
Amanita rube[illegible]s	not known
Amanita rubescens	edible
Amanita spissa	[not eaten]
Amanita vaginata	[edible]
Amanitopsis vaginata	[edible]
Amanitopsis vaginata var. *alba*	[edible]
Amanitopsis vaginata var. *plumbea*	[edible]
Amanitopsis vaginata var. *umbrinolutea*	[edible]
Armillaria mellea	edible
Armillaria ostoyae	not known
Boletus aereus	[edible]
Boletus bulbosus	not known
Boletus caudicinus	not known
Boletus communis	not known
Boletus crassus	not known
Boletus cyanescens	not known
Boletus edulis	edible

Boletus elegans	[edible]
Boletus erythropus	[edible]
Boletus esculentus	not known
Boletus leucophaeus	not known
Boletus luridus	[edible]
Boletus miniatoporus	not known
Boletus purpureus	not known
Boletus regius	[edible]
Boletus rhodoxanthus	not known
Boletus rufus	not known
Boletus scaber	[edible]
Boletus subtomentosus	[edible]
Boletus sulphureus	not known
Boletus tuberosus	not known
Boletus versipellis	not known
Bovista gigantea	not known
Bovista nigrescens	not known
Calocybe gambosa	edible
Calvatia caelata	[edible]
Calvatia maxima	not known
Calvatia utriformis	[edible]
Camarophyllus pratensis	[edible]
Cantharellus cibarius	edible
Cantharellus clavatus	not known
Cantharellus infundibuliformis	[edible]
Cantharellus tubiformis	edible
Chroogomphus rutilus	[edible]
Clavaria formosa	not known
Clavaria pallida	not known
Clavaria pistillaris	not known
Clavariadelphus pistillaris	edible
Clitocybe geotropa	[edible]
Clitocybe gibba	[edible]
Clitocybe infundibuliformis	[edible]
Clitocybe laccata	not known
Clitocybe maxima	not known
Clitocybe nebularis	edible
Clitocybe odora	edible
Clitocybe olearia	not known
Clitocybe phosphorea	not known
Clitocybe viridis	not known
Clitopilus prunulus	[edible]
Collybia badia	not known
Coprinus atramentarius	[edible]
Coprinus comatus	edible
Coprinus porcelanus	not known
Cortinarius praestans	[edible]
Craterellus clavatus	not known
Craterellus cornucopioides	edible
Dentinum repandum	not known
Fistulina buglossoides	not known
Fistulina hepatica	edible
Flammulina velutipes	[edible]
Gomphidius glutinosus	edible
Gomphidius viscidus	not known
Gomphus clavatus	edible
Gyromitra esculenta	[edible]
Gyroporus castaneus	[edible]
Gyroporus cyanescens	[edible]
Helvella crispa	edible
Helvella lacunosa	edible
Helvella mitra	not known
Helvella monacella	not known
Helvella nivea	not known
Helvella sulcata	not known
Hydnum repandum	edible
Hygrocybe punicea	[edible]
Hygrophorus eburneus	[edible]
Hygrophorus puniceus	not known
Hygrophorus russula	[edible]
Ixocomus bovinus	not known
Ixocomus elegans	not known
Ixocomus luteus	not known
Krombholzia aurantiaca	not known
Kuehneromyces mutabilis	[edible]
Laccaria amethystina	edible
Laccaria laccata	edible
Lactarius deliciosus	edible
Lactarius pergamenus	not known
Lactarius piperatus	edible
Lactarius torminosus	[edible]
Lactarius vellereus	edible
Lactarius volemus	edible
Laetiporus sulphureus	[edible]
Langermannia gigantea	edible
Lasiosphaera gigantea	not known
Leccinum aurantiacum	[edible]
Leccinum scabrum	edible
Lepista nuda	[edible]
Lepista personata	edible
Limacium eburneum	not known
Lycoperdon caelatum	[edible]
Lycoperdon echinatum	not known
Lycoperdon gemmatum	edible
Lycoperdon perlatum	edible
Lycoperdon pyriforme	edible
Macrolepiota procera	edible
Macrolepiota rhacodes	edible
Marasmius alliaceus	not known
Marasmius caryophylleus	edible
Marasmius oreades	edible
Marasmius scorodonius	[edible]
Morchella conica	[edible]
Morchella esculenta	edible
Morchella esculenta var. vulgaris	not known
Morchella rimosipes	not known
Nevrophyllum clavatum	not known
Phallus crispus	not known
Phlegmacium praestans	not known
Pholiota caperata	not known
Pleurotus ostreatus	edible
Pleurotus ostreatus f. salignus	[edible]
Pleurotus ostreatus var. columbinus	[edible]
Pleurotus pulmonarius	[edible]
Pluteus cervinus	edible
Polyporus confluens	not known
Polyporus ovinus	not known
Polyporus squamosus	edible
Pseudohydnum gelatinosum	[edible]
Ptychoverpa bohemica	edible

Ramaria aurea	edible
Ramaria botrytis	edible
Ramaria flava	[edible]
Ramaria mairei	[edible]
Rhodopaxillus personatus	not known
Rhodophyllus sinuatus	not known
Rozites caperatus	edible
Russula alutacea	edible
Russula cyanoxantha	edible
Russula emetica	[edible]
Russula olivacea	[edible]
Russula vesca	[edible]
Russula virescens	[edible]
Russula xerampelina	edible
Sarcodon imbricatus	edible
Scleroderma citrinum	[edible]
Scleroderma vulgare	not known
Scutiger confluens	not known
Scutiger ovinus	[edible]
Suillus bovinus	[edible]
Suillus granulatus	[edible]
Suillus grevillei	[edible]
Suillus luteus	edible
Tricholoma columbetta	not known
Tricholoma equestre	not known
Tricholoma flavovirens	edible
Tricholoma georgii	not known
Tricholoma personatum	not known
Tricholoma portentosum	edible
Tricholoma russula	not known
Tricholoma rutilans	not known
Tricholoma terreum	[edible]
Tricholoma tigrinum	not known
Tricholomopsis rutilans	[edible]
Verpa conica	[edible]
Verpa digitaliformis	not known
Xerocomus badius	[edible]
Xerocomus chrysenteron	edible
Xerocomus subtomentosus	edible

Burkina Faso
Rammeloo and Walleyn, 1993

Coprinus	edible
Phlebopus sudanicus	edible

Burundi
1. Buyck, 1994b; **2.** Walleyn and Rammeloo, 1994

Afroboletus luteolus	edible (1)
Amanita loosii	edible (1)
Amanita rubescens	edible (1)
C. cibarius var. *defibulatus*	edible (1)
Cantharellus congolensis	edible (1)
Cantharellus cyanescens	edible (1)
Cantharellus cyanoxanthus	edible (1)
Cantharellus densifolius	edible (1)
Cantharellus platyphyllus	edible (1)
Cantharellus pseudocibarius	edible (1)
Cantharellus ruber	edible (1)
C. rufopunctatus var. *ochraceus*	edible (1)
Cantharellus splendens	edible (1)
Cantharellus symoensii	edible (1)
Clavaria albiramea	edible (1)
Collybia aurea	edible (1)
Lactarius edulis	edible (1)
Lactarius inversus	edible (1)
Lactarius kabansus	edible (1)
Lentinus tuber-regium	[edible] (1)
Macrocybe spectabilis	edible (1)
Phlebopus colossus	[edible] (2)
Pleurotus cystidiosus	edible (1)
Russula cellulata	edible (1)
Russula phaeocephala	edible (1)
Suillus luteus	edible (1)
Termitomyces letestui	edible (1)
Termitomyces microcarpus	edible (1)
Termitomyces robustus	edible (1)
Termitomyces striatus	edible (1)
Termitomyces titanicus	edible (1)

Cameroon
1. Pegler and Vanhaecke, 1994; **2.** Rammeloo and Walleyn, 1993

Cantharellus pseudocibarius	edible (2)
Lepiota discipes	edible (2)
Marasmius hungo	edible (2)
Mycena aschi	edible (2)
Mycena bipindiensis	edible (2)
Termitomyces striatus	edible (1)

Canada
1. Marles *et al.*, 2000; **2.** Tedder, Mitchell and Farran, 2002; **3.** www.for.gov.bc.ca

Actinogyra muehlenbergii	food, medicinal (1)
Agaricus campestris	food (3)
Agaricus silvaticus	edible (2)
Amanita muscaria	medicinal (1)
Armillaria mellea	food (3)
Armillaria ostoyae	edible (2)
Boletus edulis	food (3)
Cantharellus cibarius	food (3)
Cantharellus formosus	edible (2)
Cantharellus infundibuliformis	edible (2)
Cantharellus subalbidus	edible (2)
Cetraria islandica	medicinal (1)
Cladina stellaris	food (1)
Craterellus cornucopioides	food (3)
Evernia mesomorpha	medicinal (1)
Fomes fomentarius	other – tinder (1)
Fomitopsis pinicola	medicinal, tinder (1)
Gyromitra esculenta	[edible] (3)
Hericium abietis	food (3)
Hericium erinaceus	[edible] (3)
Hydnum repandum	edible (2)
Hypomyces lactifluorum	food (3)
Inonotus obliquus	medicinal (1)
Ischnoderma resinosum	medicinal (1)
Laccaria laccata	edible (2)
Lactarius deliciosus	food (3)

Laetiporus sulphureus	edible (2)
Langermannia gigantea	food (3)
Lepista nuda	food (3)
Lycoperdon perlatum	food (3)
Macrolepiota rhacodes	edible (2)
Marasmius oreades	food (3)
Morchella elata	food (3)
Myriosclerotinia caricis-ampullaceae	medicinal (1)
Parmelia sulcata	medicinal (1)
Pleurocybella porrigens	edible (2)
Pleurotus ostreatus	food (3)
Polyozellus multiplex	edible (2)
Ptychoverpa bohemica	food (3)
Russula xerampelina	edible (2)
Sparassis crispa	edible (2)
Suillus cavipes	food (3)
Trametes suaveolens	medicinal, tinder (1)
Tricholoma caligatum	food (3)
Tricholoma magnivelare	edible (2)
Usnea hirta	medicinal (1)

Central African Republic
1. Rammeloo and Walleyn, 1993; **2.** Walleyn and Rammeloo, 1994

Agaricus subedulis	edible (1)
Collybia attenuata	edible (2)
Ganoderma curtisii	medicinal (2)
Lentinus araucariae	edible (1)
Lentinus brunneofloccosus	edible (1)
Lignosus sacer	medicinal (2)
Macrolepiota africana	edible (1)
Phlebopus sudanicus	edible (1)
Polyporus rhizomorphus	other – string (2)
Schizophyllum commune	edible (1)
Termitomyces clypeatus	edible (1)
Termitomyces schimperi	edible (1)
Volvariella volvacea	edible (1)
Xylaria papyrifera	medicinal (2)

Chile
1. FAO, 1998b; **2.** Minter, Cannon and Peredo, 1987; **3.** Schmeda-Hirschmann *et al.*, 1999a

Armillaria mellea	food (1)
Auricularia auricula-judae	edible (1)
Auricularia polytricha	edible (1)
Boletus loyo	food (1)
Clitocybe nebularis	food (1)
Coprinus atramentarius	edible (1)
Coprinus comatus	edible (1)
Cyttaria berteroi	[edible] (3)
Cyttaria darwinii	food (2)
Cyttaria espinosae	food (2)
Cyttaria hariotii	edible (1)
Cyttaria hookeri	edible (1)
Cyttaria johowii	[edible] (3)
Fistulina hepatica	edible (1)
Flammulina velutipes	food (1)
Gyromitra antartica	edible (1)
Gyromitra esculenta	edible (1)
Lactarius deliciosus	edible (1)
Macrolepiota procera	edible (1)
Morchella conica	edible, medicinal (1)
Pholiota edulis	edible (1)
Ramaria subaurantiaca	food (1)
Suillus luteus	food (1)
Volvariella speciosa	edible (1)

China
1. Birks, 1991; **2.** Cao, 1991; **3.** Chamberlain, 1996; **4.** Dong and Shen, 1993; **5.** Gong and Peng, 1993; **6.** Hall *et al.*, 1998a; **7.** Härkönen, 2002; **8.** He, 1991; **9.** Huang, 1989; **10.** Li, 1994; **11.** Liu, 1990; **12.** Liu and Yang, 1982; **13.** Guozhong, 2002, personal communication; **14.** Zang, 1984; **15.** Pegler and Vanhaecke, 1994; **16.** Tu, 1987; **17.** Winkler, 2002; **18.** www.zeri.org; **19.** Xiang and Han, 1987; **20.** Yang, 1990; **21.** Yang, 1992; **22.** Yang and Yang, 1992; **23.** Zang, 1988b; **24.** Zang and Petersen, 1990; **25.** Zang and Pu, 1992; **26.** Zang and Yang, 1991; **27.** Zang, 1988a; **28.** Zhuang, 1993; **29.** Zhuang and Wang, 1992

Agaricus arvensis	edible (6)
Agaricus augustus	edible (6)
Agaricus bisporus	edible (6)
Agaricus bitorquis	edible (6)
Agaricus blazei	edible (5)
Agaricus campestris	edible (6)
Agaricus gennadii	edible (23)
Agaricus silvaticus	edible (6)
Agaricus silvicola	edible (6)
Agrocybe cylindracea	edible (6)
Agrocybe salicacicola	edible (26)
Albatrellus confluens	edible (6)
Aleuria aurantia	edible (6)
Amanita caesarea	edible (6)
Amanita fulva	edible (12)
Amanita rubescens	edible (6)
Amanita vaginata	edible (6)
Armillaria mellea	[edible] (6)
Armillaria tabescens	edible (6)
Auricularia auricula-judae	edible (6)
Auricularia polytricha	edible (6)
Bankera fuligineoalba	medicinal (18)
Boletellus russellii	edible (6)
Boletinus pinetorum	edible (12)
Boletus aereus	edible (6)
Boletus citrifragrans	edible (14)
Boletus edulis	edible (17)
Boletus speciosus	edible (6)
Boletus violaceofuscus	edible (6)
Calocybe gambosa	edible (18)
Calvatia caelata	edible (6)
Calvatia lilacina	edible (6)
Cantharellus cibarius	edible (6)
Catathelasma ventricosum	edible (14)
Chroogomphus rutilus	edible (6)
Clavaria purpurea	edible (6)
Clitocybe clavipes	edible (6)
Clitocybe geotropa	edible (6)

Clitocybe nebularis	edible (6)	*Leucopaxillus giganteus*	edible (6)
Clitopilus prunulus	edible (6)	*Lobaria* sp.	food (7)
Collybia radicata	edible (12)	*Lycoperdon perlatum*	edible (6)
Coprinus atramentarius	edible (6)	*Lycoperdon pyriforme*	edible (6)
Coprinus cinereus	edible (6)	*Lyophyllum decastes*	edible (6)
Coprinus comatus	edible (6)	*Lyophyllum sykosporum*	edible (6)
Coprinus micaceus	edible (6)	*Macrolepiota procera*	edible (6)
Cordyceps militaris	medicinal (5)	*Macrolepiota rhacodes*	edible (6)
Cordyceps sinensis	edible (6)	*Marasmius androsaceus*	medicinal (18)
Cortinarius claricolor var. *turmalis*	edible (6)	*Marasmius oreades*	edible (6)
		Morchella conica var. rigida	edible (6)
Cortinarius collinitus	edible (6)	*Morchella crassipes*	edible (6)
Cortinarius elatior	edible (6)	*Morchella deliciosa*	edible (6)
Cortinarius praestans	edible (6)	*Morchella elata*	edible (6)
Cortinarius purpurascens	edible (6)	*Morchella esculenta*	edible (6)
Cortinarius rufo-olivaceus	food (3)	*M. esculenta* var. *rotunda*	edible (6)
Craterellus cornucopioides	edible (12)	*M. esculenta* var. *umbrina*	edible (6)
Cryptoporus volvatus	medicinal (14)	*M. esculenta* var. *vulgaris*	edible (6)
Dictyophora echinovolvata	edible (6)	*Neolentinus adhaerens*	edible (14)
Endophallus yunnanensis	edible (24)	*Neolentinus lepideus*	edible (6)
Fistulina hepatica	edible (6)	*Omphalia lapidescens*	medicinal (18)
Flammulina velutipes	edible (6)	*Oudemansiella mucida*	edible (6)
Fomes fomentarius	medicinal (18)	*Paecilomyces sinensis*	medicinal (10)
Ganoderma applanatum	medicinal (11)	*Panellus serotinus*	edible (6)
Ganoderma lucidum	edible (6)	*Phaeolepiota aurea*	edible (6)
Ganoderma sinense	medicinal (18)	*Phallus fragrans*	edible (14)
Ganoderma tsugae	edible (6)	*Phallus impudicus*	edible (6)
Gastrodia elata	edible (29)	*Phellinus baumii*	medicinal (7)
Grifola frondosa	edible (6)	*Pholiota adiposa*	edible (6)
Hericium clathroides	edible (6)	*Pholiota aurivella*	edible (6)
Hericium coralloides	edible (14)	*Pholiota nameko*	edible (6)
Hericium erinaceum	edible (14)	*Pholiota squarrosa*	edible (6)
Hericium ramosum	edible (14)	*Pleurotus abalonus*	edible (6)
Hydnum repandum	edible (6)	*Pleurotus citrinopileatus*	edible (6)
Hygrophorus arbustivus	edible (6)	*Pleurotus cornucopiae*	food (3)
Hygrophorus russula	edible (6)	*Pleurotus eryngii* var. *ferulae*	edible (18)
Hypsizygus marmoreus	edible (22)	*Pleurotus floridanus*	edible (20)
Kuehneromyces mutabilis	edible (6)	*Pleurotus ostreatus*	edible (6)
Laccaria laccata	food (3)	*Pleurotus pulmonarius*	edible (14)
Laccocephalum mylittae	edible (6)	*Pleurotus sapidus*	edible (14)
Lactarius akahatsu	edible (6)	*Polyozellus multiplex*	edible (21)
Lactarius camphoratus	edible (4)	*Polyporus cristatus*	not edible (12)
Lactarius deliciosus	edible (6)	*Polyporus squamosus*	edible (6)
Lactarius hatsudake	edible (6)	*Polyporus tubaeformis*	medicinal (7)
Lactarius quietus	edible (6)	*Polyporus umbellatus*	edible (6)
Lactarius sanguifluus	edible (6)	*Polystictus unicolor*	medicinal (18)
Lactarius subindigo	food (7)	*Psathyrella candolleana*	edible (6)
Lactarius volemus	edible (6)	*Pseudohydnum gelatinosum*	edible (14)
Laetiporus sulphureus	edible (14)	*Ptychoverpa bohemica*	edible (8)
Langermannia gigantea	edible (11)	*Ramalina* sp.	food (7)
Laricifomes officinalis	edible (6)	*Ramaria botrytis*	edible (6)
Leccinum scabrum	edible (6)	*Ramaria flavobrunnescens*	edible (12)
Lentinula edodes	edible (14)	*Ramaria obtusissima*	edible (6)
Lentinus javanicus	edible (14)	*Ramaria stricta*	food, medicinal (3)
Lentinus sajor-caju	edible (16)	*Rhizopogon piceus*	edible (9)
Lepista caespitosa	edible (6)	*Rhizopogon rubescens*	edible (6)
Lepista irina	edible (6)	*Rhodophyllus clypeatus*	edible (6)
Lepista luscina	edible (6)	*Rhodophyllus crassipes*	edible (6)
Lepista nuda	edible (6)	*Rozites caperatus*	edible (6)
Lepista personata	edible (6)	*Russula alutacea*	edible (6)
Lepista sordida	edible (6)	*Russula cyanoxantha*	edible (6)

Russula delica	edible (6)
Russula depallens	not known (12)
Russula pectinata	not known (12)
Russula rubra	edible (12)
Russula vesca	edible (6)
Russula virescens	edible (12)
Sarcodon aspratus	edible (6)
Sarcodon imbricatus	edible (14)
Schizophyllum commune	edible (6)
Scleroderma sp.	medicinal (12)
Shiraia bambusicola	medicinal (18)
Sparassis crispa	edible (6)
Sporisorium cruentum	food (13)
Suillus bovinus	edible (6)
Suillus granulatus	edible (6)
Suillus grevillei	edible (6)
Suillus luteus	edible (6)
Termitomyces albuminosus	edible (27)
Termitomyces clypeatus	edible (6)
Termitomyces cylindricus	edible (15)
Termitomyces eurhizus	edible (15)
Termitomyces heimii	edible (20)
Termitomyces microcarpus	edible (15)
Thamnolia vermicularis	[food] (3)
Thelephora ganbajun	food (3)
Thelephora vialis	not edible (12)
Trametes robiniophila	edible (28)
Trametes sanguinea	medicinal (18)
Trametes versicolor	edible (6)
Tremella aurantia	edible (6)
Tremella foliacea	edible (6)
Tremella fuciformis	[medicinal] (3)
Tremella lutescens	edible (14)
Tremella mesenterica	edible (6)
Tricholoma bakamatsutake	edible (17)
Tricholoma caligatum	edible (6)
Tricholoma flavovirens	edible (6)
Tricholoma magnivelare	edible (6)
Tricholoma matsutake	edible (17)
Tricholoma mongolicum	edible (11)
Tricholoma portentosum	edible (6)
Tricholoma quercicola	edible (17)
Tricholoma saponaceum	edible (6)
Tricholoma sejunctum	edible (6)
Tricholoma terreum	edible (6)
Tuber aestivum	edible (6)
Tuber brumale	edible (6)
Tuber indicum	edible (25)
Tuber melanosporum	edible (6)
Tuber rufum	edible (6)
Tuber sinosum	edible (6)
Tyromyces sulphureus	medicinal (18)
Umbilicaria esculenta	edible (6)
Usnea sp.	medicinal (1)
Volvariella bombycina	edible (6)
Volvariella esculenta	edible (19)
Volvariella volvacea	edible (6)
Wolfiporia extensa	edible (6)
Wynnella silvicola	edible (2)

Congo [Democratic Republic of the]
1. Degreef *et al.*, 1997; **2.** Pegler and Vanhaecke, 1994; **3.** Rammeloo and Walleyn, 1993; **4.** Walleyn and Rammeloo, 1994

Agaricus erythrotrichus	edible (3)
Agaricus goossensiae	edible (3)
Agaricus nivescens	edible (3)
Agaricus volvatulus	edible (1)
Amanita aurea	edible (1)
Amanita goosensiae	edible (3)
Amanita loosii	edible (1)
Amanita robusta	edible (1)
Amanita zambiana	edible (3)
Amanitopsis pudica	edible (4)
Auricularia auricula-judae	edible (3)
Auricularia delicata	edible (3)
Auricularia polytricha	edible (3)
Auricularia tenuis	edible (1)
Bondarzewia berkeleyi	edible (3)
Camarophyllus subpratensis	edible (3)
Cantharellus cibarius	edible (3)
C. cibarius var. *defibulatus*	edible (1)
C. cibarius var. *latifolius*	edible (1)
Cantharellus congolensis	edible (1)
Cantharellus cyanoxanthus	edible (3)
Cantharellus densifolius	edible (1)
Cantharellus incarnatus	edible (1)
Cantharellus luteopunctatus	edible (1)
Cantharellus miniatescens	edible (1)
Cantharellus platyphyllus	edible (1)
Cantharellus pseudofriesii	edible (3)
Cantharellus ruber	edible (1)
Cantharellus rufopunctatus	edible (1)
Cantharellus symoensii	edible (1)
Clavaria albiramea	edible (1)
Collybia piperata	edible (3)
Cookeina sulcipes	edible (3)
Corditubera bovonei	edible (4)
Cotylidia aurantiaca	edible (1)
Craterellus aureus	edible (3)
C. cornucopioides var. *cornucopioides*	edible (1)
C. cornucopioides var. *parvisporus*	edible (3)
Cymatoderma elegans subsp. *infundibuliforme*	edible (1)
Goossensia cibarioides	edible (3)
Gymnopilus sp.	edible (3)
Hypholoma wambensis	edible (3)
Lactarius angustus	edible (3)
Lactarius congolensis	edible (4)
Lactarius edulis	edible (1)
Lactarius inversus	edible (1)
Lactarius kabansus	edible (1)
Lactarius latifolius	edible (1)
Lactarius pelliculatus f. *pallidus*	edible (3)
Lactarius pseudovolemus	edible (1)
Lactarius sesemotani	edible (3)
Lentinus citrinus	edible (4)
Lentinus sajor-caju	edible (3)
Lentinus squarrulosus	edible (3)

Lentinus tuber-regium	edible (3)
Lentinus velutinus	edible (4)
Lenzites elegans	medicinal, [poisonous] (4)
Lepiota henningsii	edible (3)
Leucoagaricus bisporus	edible (3)
Leucocoprinus discoideus	edible (3)
Macrolepiota africana	edible (3)
M. gracilenta var. *goossensiae*	edible (1)
Macrolepiota procera	edible (1)
Macrolepiota prominens	edible (3)
Macrolepiota zeyheri	edible (3)
Marasmius arborescens	edible (4)
Marasmius buzungolo	edible (3)
Marasmius crinis-equi	other – jewelry (4)
Marasmius grandisetulosus	edible (4)
Marasmius piperodora	edible (4)
Nothopanus hygrophanus	edible (3)
Oudemansiella canarii	edible (4)
Peziza vesiculosa	edible (3)
Phallus indusiatus	medicinal (4)
Phlebopus colossus	[edible] (4)
Pleurotus djamor	edible (3)
Pluteus cervinus var. *ealaensis*	edible (3)
Psathyrella spadicea	edible (1)
Pseudocraterellus laeticolor	edible (1)
Pycnoporus sanguineus	medicinal (4)
Ramaria ochracea	edible (4)
Russula atrovirens	edible (3)
Russula cellulata	edible (1)
Russula cyclosperma	edible (3)
Russula diffusa var. *diffusa*	edible (1)
Russula hiemisilvae	edible (1)
Russula pseudostriatoviridis	edible (3)
Russula roseoalba	edible (3)
Russula roseostriata	edible (3)
Russula sese	edible (3)
Russula sesenagula	edible (3)
Russula striatoviridis	edible (3)
Schizophyllum commune	edible (1)
Scleroderma bovonei	edible (3)
Termitomyces aurantiacus	edible (3)
Termitomyces clypeatus	edible (3)
Termitomyces entolomoides	edible (2)
Termitomyces globulus	edible (3)
Termitomyces letestui	edible (1)
Termitomyces mammiformis	edible (3)
Termitomyces microcarpus	edible (1)
Termitomyces robustus	edible (3)
Termitomyces schimperi	edible (1)
Termitomyces striatus	edible (2)
T. striatus var. *aurantiacus*	edible (1)
Termitomyces titanicus	edible (3)
Trogia infundibuliformis	edible (3)
Volvariella bombycina	edible (3)
Volvariella parvispora	edible (3)
Xerocomus pallidosporus	edible (3)

Congo [Republic of the]
Rammeloo and Walleyn, 1993

Armillaria distans	edible
Chlorophyllum molybdites	edible
Collybia anombe	edible
Collybia oronga	edible
Leucocoprinus gandour	edible
Phlebopus sudanicus	edible

Costa Rica
Saenz, Lizano and Nassar, 1983

Agaricus silvaticus	edible
Amanita caesarea	edible
Amanita gemmata	edible
Amanita inaurata	edible
Amanita muscaria	hallucinogen, poisonous
Amanita rubescens	edible
Amanita vaginata	edible
Armillaria mellea	edible
Armillaria tabescens	edible
Aspropaxillus lepistoides	edible
Cantharellus cibarius	edible
Collybia distorta	edible
Collybia dryophila	edible
Collybia familia	edible
Collybia platyphylla	edible
Coprinus comatus	edible
Cortinarius purpurascens	edible
Craterellus cornucopioides	edible
Fistulina hepatica	edible
Helvella lacunosa	edible
Hericium erinaceus	edible
Hydnum umbilicatum	edible
Hygrophorus penarius	edible
Laccaria amethystina	edible
Laccaria laccata	edible
Lacrymaria velutina	edible
Lactarius deliciosus	edible
Lactarius indigo	edible
Lactarius mitissimus	edible
Lactarius vellereus	edible
Lepista nuda	edible
Leucopaxillus giganteus	edible
Lyophyllum aggregatum	edible
Macrolepiota gracilenta	edible
Macrolepiota procera	edible
Marasmius oreades	edible
Melanoleuca grammopodia	edible
Morchella esculenta	edible
Morchella esculenta var. *rotunda*	edible
Mycena pura	edible
Panaeolus cyanescens	hallucinogen
Pleurotus ostreatus	edible
Pleurotus salignus	edible
Pluteus cervinus	edible
Psilocybe aztecorum	hallucinogen
Psilocybe cubensis	hallucinogen
Psilocybe mexicana	hallucinogen
Rhodophyllus aprilis	edible
Russula atropurpurea	edible
Russula chamaeleontina	edible
Russula cyanoxantha	edible

Russula erythropus	edible
Russula lepida	edible
Russula minutula	edible
Russula viscida	edible
Volvariella bakeri	edible
Volvariella bombycina	edible
Volvariella speciosa	edible
Xerula radicata	edible

Cote d'Ivoire
1. Ducousso, Ba and Thoen, 2002; **2.** Locquin, 1954; **3.** Pegler and Vanhaecke, 1994; **4.** Rammeloo and Walleyn, 1993

Hygrophoropsis aurantiaca	edible (4)
Hygrophoropsis mangenotii	edible (2)
Russula sp.	food (1)
Termitomyces striatus	edible (3)

Egypt
Zakhary *et al.*, 1983

Agaricus campestris	edible
Agaricus rodmani	edible
Collybia sp.	edible

Ethiopia
Tuno, 2001

Lentinus sp.	edible
Schizophyllum commune	edible

Fiji
Markham, 1998

Auricularia sp.	food

Gabon
1. Rammeloo and Walleyn, 1993; **2.** Walleyn and Rammeloo, 1994. Note: another 15+ types are listed in Walker, 1931 by local name only

Cantharellus sp.	edible (1)
Daldinia sp.	medicinal (2)
Lentinus tuber-regium	edible (1)
Polyporus rhizomorphus	other – string (2)
Pycnoporus sp.	medicinal (2)

Ghana
1. Ducousso, Ba and Thoen, 2002; **2.** Obodai and Apetorgbor, 2001; **3.** Rammeloo and Walleyn, 1993; **4.** Walleyn and Rammeloo, 1994

Agaricus campestris	edible (3)
Agaricus goossensiae	edible (3)
Auricularia sp.	edible, medicinal (2)
Calvatia excipuliformis	medicinal (2)
Cantharellus floridulus	food (1)
Coprinus micaceus	edible (2)
Daldinia concentrica	medicinal (2)
Ganoderma lucidum	medicinal (2)
Lentinus tuber-regium	medicinal (2)
Macrolepiota procera	edible (3)
Mycena flavescens	edible (2)
Phlebopus colossus	[edible] (4)
Pluteus subcervinus	edible (2)
Psathyrella sp.	edible (2)
Schizophyllum commune	edible, medicinal (2)
Termitomyces sp.	edible (2)
Volvariella volvacea	edible (3)

Greece
Diamandis, 2002, personal communication

Agaricus arvensis	food
Agaricus campestris	food
Amanita caesarea	food
Boletus spp.	food
Cantharellus cibarius	food
Coprinus sp.	food
Lactarius deliciosus	food
Macrolepiota sp.	food
Pleurotus ostreatus	food
Ramaria sp.	food

Guatemala
Flores, 2002, personal communication

Agaricus campestris	food
Agrocybe aegerita	food
Amanita caesarea	food
Amanita calyptroderma	food
Amanita hemibapha	food
Amanita rubescens	food
Armillaria mellea	food
Auricularia delicata	food
Boletus edulis	food
Cantharellus cibarius	food
Cantharellus ignicolor	food
Cantharellus odoratus	food
Catathelasma ventricosum	food
Clavulina cinerea	food
Cortinarius praestans	food
Helvella crispa	food
Helvella lacunosa	food
Hydnum repandum	food
Hygrophorus russula	food
Hypomyces lactifluorum	food
Laccaria amethystea	food
Laccaria bicolor	food
Laccaria laccata	food
Lactarius corrugis	food
Lactarius deliciosus	food
Lactarius indigo	food
Lactarius rubrilacteus	food
Lactarius salmonicolor	food
Morchella esculenta	food
Pleurotus sp.	food
Ramaria araiospora	food
Ramaria botrytis	food
Ramaria flava	food
Russula delica	food
Schizophyllum commune	food
Tremella reticulata	food

Tricholoma flavovirens	food
Trogia sp.	food

GUINEA
Walleyn and Rammeloo, 1994

Lepiota grassei	edible

GUYANA
Simmons, Henkel and Bas, 2002

Amanita perphaea	edible

HONG KONG SPECIAL ADMINISTRATIVE REGION, CHINA
Chang and Mao, 1995

Agaricus abruptibulbus	edible
Agaricus arvensis	edible, medicinal
Agaricus bisporus	edible, medicinal
Agaricus campestris	edible, medicinal
Agaricus comtulus	edible
Agaricus micromegethus	edible
Agaricus placomyces	[edible]
Agaricus purpurellus	[edible]
Agaricus rubellus	edible
Agaricus semotus	[edible]
Agaricus silvaticus	edible
Agaricus silvicola	edible
Agrocybe cylindracea	edible, medicinal
Agrocybe dura	edible, medicinal
Agrocybe farinacea	edible
Agrocybe paludosa	edible
Agrocybe pediades	edible
Agrocybe praecox	edible, medicinal
Amanita rubescens	edible
Amanita vaginata	[edible]
Amanita virgineoides	edible
Amauroderma nigrum	medicinal
Amauroderma rude	medicinal
Armillaria mellea	edible, medicinal
Astraeus hygrometricus	medicinal
Auricularia auricula-judae	edible, medicinal
Auricularia delicata	edible, medicinal
Auricularia fuscosuccinea	edible
Auricularia polytricha	edible, medicinal
Boletus emodensis	edible
Boletus griseus	edible
Boletus speciosus	[edible]
Bovista plumbea	edible, medicinal
Bovistella sinensis	medicinal
Calocera cornea	edible
Calocera viscosa	edible
Calocybe leucocephala	edible
Calvatia caelata	medicinal
Calvatia craniiformis	edible, medicinal
Calvatia cyathiformis	edible, medicinal
Calvatia lilacina	edible, medicinal
Camarophyllus virgineus	edible
Cantharellus cinereus	edible
Cantharellus cinnabarinus	edible
Cerrena unicolor	medicinal
Clavaria vermicularis	edible
Clavicorona pyxidata	edible
Clavulina cristata	edible
Clavulinopsis helvola	edible
Clavulinopsis miyabeana	edible
Clitocybe clavipes	edible
Clitocybe fragrans	[edible], medicinal
Clitopilus prunulus	edible
Collybia acervata	[edible]
Collybia butyracea	[edible]
Collybia confluens	edible
Collybia dryophila	[edible]
Coprinus comatus	[edible], medicinal
Coprinus micaceus	[edible], medicinal
Coprinus plicatilis	edible, medicinal
Coprinus radians	edible, medicinal
Coprinus sterquilinus	edible, medicinal
Craterellus aureus	edible
Craterellus cornucopioides	edible
Crepidotus applanatus	edible
Crepidotus mollis	edible
Cyathus stercoreus	medicinal
Cystoderma amianthinum	edible
Cystoderma terrei	edible
Dacrymyces palmatus	edible
Dacryopinax spathularia	edible
Dictyophora duplicata	edible, medicinal
D. indusiata f. lutea	edible
Dictyophora multicolor	[medicinal]
Flammulina velutipes	edible
Fomes fomentarius	medicinal
Fomitopsis ulmaria	medicinal
Ganoderma applanatum	medicinal
Ganoderma capense	medicinal
Ganoderma lobatum	medicinal
Ganoderma lucidum	medicinal
Ganoderma sinense	medicinal
Ganoderma tenue	medicinal
Ganoderma tropicum	medicinal
Ganoderma tsugae	medicinal
Geastrum triplex	medicinal
Gomphus clavatus	edible
Grifola frondosa	edible, medicinal
Gyrodon lividus	edible
Gyroporus castaneus	[edible]
Hericium erinaceus	edible, medicinal
Hexagonia apiaria	medicinal
Hirschioporus abietinum	medicinal
Hirschioporus fuscoviolaceum	medicinal
Hohenbuehelia petaloides	edible
Hygrocybe cantharellus	edible
Hygrophoropsis aurantiaca	edible
Hygrophorus eburneus	edible
Hypsizygus marmoreus	edible
Ischnoderma resinosum	medicinal
Kobayasia nipponica	edible
Kuehneromyces mutabilis	edible
Laccaria amethystea	edible
Laccaria laccata	edible
Lacrymaria velutina	edible
Lactarius akahatsu	edible
Lactarius deliciosus	edible

Lactarius hatsudake	edible, medicinal
Lactarius hygrophoroides	edible, medicinal
Lactarius volemus	edible, medicinal
Langermannia gigantea	edible, medicinal
Lanopila nipponica	edible
Lentinellus cochleatus	edible
Lentinula edodes	edible, medicinal
Lentinus sajor-caju	edible
Lentinus strigosus	edible
Lentinus tigrinus	edible
Lepiota aspera	edible
Lepiota clypeolaria	[edible]
Lepiota ventriosospora	edible
Lepista nuda	edible
Lepista sordida	edible
Leucoagaricus pudicus	[edible]
Leucocoprinus cepaestipes	[edible], medicinal
Limacella glioderma	edible
Lycoperdon asperum	medicinal
Lycoperdon perlatum	edible, medicinal
Lycoperdon pusillum	medicinal
Lycoperdon pyriforme	edible, medicinal
Lycoperdon spadiceum	medicinal
Lyophyllum fumosum	edible
Lyophyllum ulmarium	edible
Lysurus mokusin	medicinal
Macrocybe lobayensis	edible
Macrolepiota procera	edible, medicinal
Macrolepiota rhacodes	edible, medicinal
Marasmiellus ramealis	edible, medicinal
Marasmius cohaerens	edible, medicinal
Marasmius crinis-equi	edible
Marasmius maximus	edible
Marasmius oreades	edible, medicinal
Marasmius personatus	edible
Marasmius purpureostriatus	edible
Megacollybia platyphylla	edible, medicinal
Melanoleuca alboflavida	edible
Melanoleuca melaleuca	edible
Mycena pura	edible
Neolentinus lepideus	edible, medicinal
Ossicaulis lignatilis	edible
Oudemansiella mucida	edible, medicinal
Panellus serotinus	edible
Panellus stipticus	[medicinal]
Phallus impudicus	edible, medicinal
Phallus rubicundus	[medicinal]
Phallus tenuis	medicinal
Phellinus conchatus	medicinal
Phellinus igniarius	medicinal
Pholiota flammans	edible, medicinal
Pholiota highlandensis	edible
Pholiota nameko	edible, medicinal
Phylloporus rhodoxanthus	edible
Pisolithus tinctorius	medicinal
Pleurocybella porrigens	edible
Pleurotus citrinopileatus	edible, medicinal
Pleurotus cornucopiae	edible
Pleurotus corticatus	edible, medicinal
Pleurotus cystidiosus	edible
Pleurotus flexilis	edible
Pleurotus ostreatus	edible, medicinal
Pleurotus pulmonarius	edible
Pleurotus rhodophyllus	edible
Pleurotus spodoleucus	edible
Pluteus leoninus	edible
Pluteus pellitus	edible
Pluteus tricuspidatus	edible
Polyporus alveolaris	medicinal
Polyporus arcularius	edible, medicinal
Polyporus elegans	medicinal
Polyporus umbellatus	edible, medicinal
Psathyrella candolleana	edible
Psathyrella piluliformis	edible
Psathyrella rugocephala	edible
Pulveroboletus ravenelii	[edible], medicinal
Pycnoporus cinnabarinus	medicinal
Pycnoporus coccineus	medicinal
Ramaria apiculata	edible
Rhizopogon rubescens	edible
Russula aeruginea	edible
Russula crustosa	edible, medicinal
Russula delica	edible, medicinal
Russula emetica	[medicinal]
Russula foetens	[medicinal]
Russula lilacea	edible, medicinal
Russula sanguinea	edible, medicinal
Russula senecis	[medicinal]
Russula vesca	edible, medicinal
Russula virescens	edible, medicinal
Sarcoscypha coccinea	edible
Schizophyllum commune	edible
Scleroderma bovista	edible, medicinal
Scleroderma citrinum	[medicinal]
Scleroderma flavidum	medicinal
Scleroderma polyrhizum	edible, medicinal
Scleroderma verrucosum	medicinal
Strobilomyces confusus	edible
Strobilomyces strobilaceus	edible, medicinal
Stropharia coronilla	edible, medicinal
Stropharia rugosoannulata	edible
Stropharia semiglobata	edible, medicinal
Suillus americanus	edible
Suillus brevipes	edible
Suillus granulatus	edible, medicinal
Suillus lactifluus	edible
Suillus subluteus	edible
Suillus tomentosus	edible
Termitomyces albuminosus	edible, medicinal
Termitomyces clypeatus	edible
Termitomyces microcarpus	edible
Trametes albida	medicinal
Trametes hirsuta	medicinal
Trametes orientalis	medicinal
Trametes pubescens	medicinal
Trametes versicolor	medicinal
Tremella fuciformis	edible, medicinal
Tricholoma imbricatum	edible
Tricholoma pessundatum	edible
Tricholoma rutilans	[edible]
Tylopilus ballouii	edible
Volvariella bombycina	edible, medicinal

Volvariella speciosa	edible
V. speciosa var. *gloiocephala*	[edible]
Volvariella volvacea	edible, medicinal
Wolfiporia extensa	edible, medicinal
Xerocomus chrysenteron	edible
Xeromphalina campanella	edible
Xerula radicata	edible
Xylaria polymorpha	medicinal

INDIA

1. Birks, 1991; **2.** Boruah *et al.*, 1996; **3.** Singh and Rawat, 2000; **4.** Harsh, Tiwari and Rai, 1996; **5.** Pegler and Vanhaecke, 1994; **6.** Purkayastha and Chandra, 1985; **7.** Richardson, 1991; **8.** Sarkar, Chakraborty and Bhattacharjee, 1988; **9.** Sharda, Kaushal and Negi, 1997; **10.** Sharma and Doshi, 1996

Agaricus campestris	edible (6)
Amanita vaginata	edible (6)
Astraeus hygrometricus	edible (6)
Auricularia delicata	edible (6)
Boletus edulis	edible (2)
Bovista apedicellata	medicinal (4)
Bovista gigantea	edible (6)
Calocybe indica	edible (8)
Calvatia cyathiformis	edible (4)
Cantharellus cibarius	edible (6)
Cantharellus floccosus	edible (2)
Cetrariastrum sp.	other – spice (7)
Clavaria aurea	edible (6)
Clitocybe sp.	edible (6)
Collybia sp.	edible (6)
Coprinus acuminatus	edible (6)
Coprinus atramentarius	edible (6)
Coprinus comatus	edible (6)
Cyathus limbatus	medicinal (4)
Daldinia concentrica	medicinal (4)
Entoloma microcarpum	edible (6)
Evernia prunastri	other – perfume (7)
Ganoderma lucidum	medicinal (4)
Geastrum fimbriatum	edible (4)
Geastrum triplex	edible (4)
Geopora sp.	edible (6)
Helvella sp.	edible (6)
Lactarius deterrimus	edible (6)
Lactarius princeps	edible (6)
Langermannia gigantea	edible (6)
Lentinula edodes	edible (6)
Lentinus sajor-caju	edible (6)
Lentinus subnudus	edible (6)
Lepiota mastoidea	edible (6)
Limacella sp.	edible (6)
Lycoperdon pusillum	edible (4)
Lycoperdon pyriforme	edible (6)
Macrocybe gigantea	edible (6)
Macrocybe lobayensis	edible (8)
Macrolepiota procera	edible (6)
Marasmius sp.	edible (8)
Microporus xanthopus	medicinal (4)
Morchella angusticeps	edible (3)
Morchella esculenta	edible (3)
Mycenastrum corium	edible (4)
Parmotrema sp.	other – spice (7)
Peltigera canina	medicinal (1)
Phallus impudicus	medicinal (4)
Phellorinia inquinans	edible (10)
Pleurotus eryngii	edible (6)
Pleurotus fossulatus	edible (6)
Pleurotus ostreatus	edible (6)
Podabrella microcarpa	edible (4)
Podaxis pistillaris	edible (6)
Podoscypha nitidula	edible (6)
R. flavobrunnescens var. *aurea*	food (9)
R. flavobrunnescens var. *formosoides*	[edible] (9)
R. flavobrunnescens var. *typica*	food (9)
Ramaria obtusissima	food (9)
Ramaria sandaracina	[edible] (9)
Ramaria sanguinea	food (6)
Ramaria subbotrytis	food (9)
Russula delica	edible (6)
Russula densifolia	edible (6)
Schizophyllum commune	edible (6)
Scleroderma radicans	edible (4)
Scleroderma verrucosum	edible (6)
Sparassis crispa	edible (6)
Termitomyces albuminosus	edible (6)
Termitomyces clypeatus	edible (5)
Termitomyces eurhizus	edible (8)
Termitomyces heimii	edible (4)
Termitomyces microcarpus	edible (8)
Termitomyces radicatus	edible (5)
Termitomyces striatus	edible (5)
Tricholoma sulphureum	food (6)
Tuber sp.	edible (6)
Volvariella diplasia	edible (8)
Volvariella terastria	edible (6)
Volvariella volvacea	edible (8)
Xylaria polymorpha	medicinal (4)

INDONESIA

1. Burkhill, 1935; **2.** Ducousso, Ba and Thoen, 2002

Auricularia auricula-judae	food, medicinal (1)
Clitocybe hypocalamus	food (1)
Marasmius sp.	food (1)
Polyporus grammocephalus	food (1)
Russula sp.	edible (1)
Scleroderma sp.	food (2)
Termitomyces albuminosus	food (1)

IRAQ

1. Al-Naama, Ewaze and Nema, 1988; **2.** Alsheikh and Trappe, 1983

Terfezia claveryi	edible (1)
Tirmania nivea	edible (2)
Tirmania pinoyi	edible (2)

Israel Wasser, 1995	
Pleurotus eryngii var. *ferulae*	edible
Suillus granulatus	edible
Volvariella speciosa	edible

Jordan 1. Ereifej and Al-Raddad, 2000; 2. Sabra and Walter, 2001	
Agaricus campestris	food (2)
Boletus aestivalis	food (2)
Boletus erythropus	edible (1)
Cantharellus cibarius	food (2)
Entoloma clypeatum	edible (1)
Lactarius deliciosus	food (2)
Lepista nuda	food (2)
Lyophyllum decastes	food (2)
Pleurotus eryngii	food (2)

Kenya 1. Pegler and Vanhaecke, 1994; 2. Rammeloo and Walleyn, 1993; 3. Walleyn and Rammeloo, 1994	
Agaricus campestris	edible (2)
Coprinus sterquilinus	edible (3)
Engleromyces goetzei	medicinal (3)
Langermannia wahlbergii	other – dye (3)
Lignosus sacer	medicinal (3)
Macrolepiota dolichaula	edible (2)
Phlebopus sudanicus	[hallucinogen] (3)
Podaxis pistillaris	other – dye (3)
Psilocybe merdaria	hallucinogen, poisonous (3)
Termitomyces eurhizus	edible (2)
Termitomyces striatus	edible (1)

Korea [Democratic People's Republic of] Wang, Hall and Evans, 1997	
Tricholoma matsutake	edible

Kuwait Alsheikh and Trappe, 1983	
Tirmania nivea	edible
Tirmania pinoyi	food, medicinal

Kyrgyzstan El'chibaev, 1964	
Armillaria mellea	edible
Bovista plumbea	edible
Calvatia caelata	edible
Cantharellus cibarius	edible
Coprinus atramentarius	edible
Coprinus comatus	[edible]
Flammulina velutipes	edible
Gyromitra esculenta	[edible]
Lactarius deliciosus	edible
Laetiporus sulphureus	edible
Leccinum scabra	edible
Lepista nuda	edible
Lycoperdon gemmatum	edible
Lycoperdon pyriforme	edible
Macrolepiota excoriata	edible
Morchella conica	edible
Morchella intermedia	edible
Paxillus atrotomentosus	edible
Pleurotus eryngii	edible
Pleurotus ostreatus	edible
Polyporus squamosus	edible
Ptychoverpa bohemica	edible
Ramaria flava	edible
Russula aeruginea	edible
Russula delica	edible
Russula nitida	edible
Russula olivascens	edible
Russula rosacea	edible
Russula sardonia	edible
Sarcodon imbricatus	edible
Scleroderma citrinum	[edible]
Tricholoma portentosum	edible

Lao People's Democratic Republic 1. Hosaka, 2002, personal communication; 2. http//giechgroup.hp.infoseek.co.jp/kinoko/eng.html	
Amanita hemibapha	food (1)
Amanita vaginata	[edible] (2)
Amanita virgineoides	[edible] (2)
Armillaria sp.	[edible] (2)
Cantharellus luteocomus	[edible] (2)
Collybia acervata	[edible] (2)
Coprinus disseminatus	[edible] (2)
Ganoderma lucidum	[medicinal] (2)
Hygrocybe cantharellus	[edible] (2)
Hygrocybe conica	[edible] (2)
Hygrocybe punicea	[edible] (2)
Laccaria amethystea	[edible] (2)
Laccaria laccata	[edible] (2)
Lactarius volemus	[edible] (2)
Leccinum extremiorientale	[edible] (2)
Lentinus strigosus	food (1)
Lyophyllum sp.	[edible] (2)
Macrolepiota procera	[edible] (2)
Oudemansiella sp.	[edible] (2)
Pycnoporus coccineus	[other] (2)
Ramaria	medicinal (1)
Russula	food (1)
Russula densifolia	[edible] (2)
Russula virescens	[edible] (2)
Schizophyllum commune	food (1)
Suillus granulatus	[edible] (2)
Termitomyces	food (1)
Trametes versicolor	[medicinal] (2)

Lesotho Rammeloo and Walleyn, 1993	
Termitomyces	edible

LIBYAN ARAB JAMAHIRIYA
Alsheikh and Trappe, 1983

Tirmania nivea	edible
Tirmania pinoyi	edible

MADAGASCAR
1. Bouriquet, 1970; **2.** Ducousso, Ba and Thoen, 2002; **3.** Rammeloo and Walleyn, 1993; **4.** Richardson, 1991; **5.** Walleyn and Rammeloo, 1994

Agaricus campestris	edible (5)
Agaricus silvicola	[edible] (1)
Amanita hovae	edible (3)
Amanita robusta	[edible] (1)
Amanita vaginata	[edible] (1)
Anthurus pentulus	[edible] (1)
Armillaria heimii	[edible] (1)
Armillariella elegans	[edible] (1)
Aseroë sp.	[edible] (1)
Auricularia auricula-judae	[edible] (1)
Boletus bouriqueti	[edible] (1)
Boletus colossus	[edible] (1)
Cantharellus cibarius	[edible] (1)
Cantharellus cyanoxanthus	[edible] (1)
Cantharellus eucalyptorum	food (2)
Cantharellus madagascariensis	[edible] (1)
Chlorophyllum madagascariense	edible (3)
Chlorophyllum molybdites	edible (3)
Clathrus madagascariensis	[edible] (1)
Clavaria miniata	[edible] (1)
Collybia tamatavae	edible (3)
Cortinarius largus	edible (5)
Cyathus stercoreus	[edible] (1)
Galiella javanica	medicinal (5)
Ganoderma lucidum	[edible] (1)
Geastrum fimbriatum	[edible] (1)
Inocybe	[edible] (1)
Inocybe tulearensis	[edible] (5)
Laccaria edulis	edible (3)
Lactarius rubroviolascens	[edible] (1)
Lentinus berteri	[edible] (1)
Lentinus tuber-regium	edible (3); medicinal (5)
Lenzites palisoti	[edible] (1)
Lepiota aspera	[edible] (1)
Lepiota imerinensis	[edible] (1)
Lepiota madagascariensis	[edible] (1)
Lepiota madirokelensis	edible (3)
Lepiota rabarijanonae	[edible] (5)
Lepiota roseoalba	[edible] (5)
Leucocoprinus badhamii	[edible] (1)
Leucocoprinus imerinensis	edible (3)
Leucocoprinus nanianae	edible (5)
Leucocoprinus tanetensis	edible (3)
Lycoperdon endotephrum	edible (5)
Lysurus periphragmoides	[edible] (1)
Macrocybe spectabilis	edible (3)
Macrolepiota excoriata	[edible] (1)
M. excoriata var. *rubescens*	edible (3)
Macrolepiota procera	[edible] (1)
M. procera var. *vezo*	edible (5)
Microporus sanguineus	[edible] (1)
Morchella intermedia	edible (3)
Mutinus bambusinus	[edible] (1)
Phaeolus manihotis	[edible] (1)
Phallus armeniacus	[edible] (1)
Phallus impudicus	[edible] (1)
Phlebopus colossus	edible (3)
Pleurotus dactylophorus	[edible] (1)
Podaxon termitophilus	[edible] (1)
Polyporus croceoleucus	[edible] (1)
Polystictus sp.	[edible] (1)
Ramaria stricta	[edible] (1)
Roccella sp.	other – dye (4)
Russula cyanoxantha	[edible] (1)
Russula madagassensis	edible (5)
Schizophyllum commune	edible (3)
Strobilomyces	[edible] (1)
Strobilomyces coturnix	edible (5)
Suillus granulatus	edible (3)
Terfezia decaryi	[edible] (1)
Tricholoma scabrum	edible (3)
Volvariella esculenta	[edible] (1)
Volvariella volvacea	edible (3)
Xerocomus chrysenteron	[edible] (1)
Xerocomus versicolor	edible (3)

MALAWI
1. Rammeloo and Walleyn, 1993; **2.** Walleyn and Rammeloo, 1994; see also www.malawifungi.org

Afroboletus costatisporus	edible (1)
Afroboletus luteolus	edible (1)
Agaricus bingensis	edible (1)
Agaricus campestris	edible (1)
Agaricus croceolutescens	edible (1)
Agaricus endoxanthus	edible (1)
Amanita bingensis	edible (1)
Amanita calopus	edible (1)
Amanita flammeola	edible (1)
Amanita fulva	edible (1)
Amanita goosensiae	edible (1)
Amanita hemibapha	edible (1)
Amanita muscaria	hallucinogen, poisonous, (2)
Amanita praeclara	[edible], insecticidal (2)
Amanita rhodophylla	edible (1)
Amanita robusta	edible (1)
Amanita rubescens	edible (1)
Amanita vaginata	edible (1)
Amanita zambiana	edible (1)
Auricularia auricula-judae	edible (1)
Auricularia delicata	edible (1)
Cantharellus cibarius	edible (1)
Cantharellus congolensis	edible (1)
Cantharellus longisporus	edible (1)
Cantharellus tenuis	edible (1)
Clavaria albiramea	edible (1)
Collybia confluens	edible (1)
Collybia dryophila	edible (1)
Coprinus disseminatus	edible (1)
Cymatoderma dendriticum	edible (1)

Gyroporus castaneus	edible (1)
Inocybe	[edible] (1)
Lactarius gymnocarpus	edible (1)
Lactarius piperatus	edible (1)
Lactarius vellereus	edible (1)
Lentinus cladopus	edible (1)
Lentinus squarrulosus	edible (1)
Lepista caffrorum	edible (1)
Macrocybe lobayensis	edible (1)
Macrolepiota dolichaula	edible (1)
Macrolepiota procera	edible (1)
Micropsalliota brunneosperma	edible (1)
Perenniporia mundula	medicinal (2)
Phlebopus colossus	edible (1)
Phlebopus sudanicus	edible (1)
Polyporus brasiliensis	edible (1)
Polyporus moluccensis	edible (2)
Psathyrella atroumbonata	[edible] (2)
Psathyrella candolleana	edible (1)
Pulveroboletus aberrans	edible (1)
Pycnoporus sanguineus	edible (1)
Rubinoboletus luteopurpureus	edible (1)
Russula afronigricans	edible (1)
Russula cyanoxantha	edible (1)
Russula delica	edible (1)
Russula ochroleuca	edible (1)
Russula rosea	edible (1)
Russula schizoderma	edible (1)
Schizophyllum commune	edible (1)
Stereopsis hiscens	edible (1)
Suillus granulatus	edible (1)
Suillus luteus	edible (1)
Termitomyces aurantiacus	edible (1)
Termitomyces clypeatus	edible (1)
Termitomyces eurhizus	edible (1)
Termitomyces microcarpus	edible (1)
Termitomyces robustus	edible (1)
Termitomyces schimperi	edible (1)
Termitomyces striatus	edible (1)
Termitomyces titanicus	edible (1)
Trogia infundibuliformis	[edible] (2)
Tubosaeta brunneosetosa	edible (1)
Vascellum pratense	edible (1)
Volvariella volvacea	edible (1)
Xerocomus pallidosporus	edible (1)
Xerocomus soyeri	edible (1)
Xerula radicata	edible (1)

Malaysia
1. Burkhill, 1935; **2.** Pegler and Vanhaecke, 1994

Termitomyces albuminosus	food (1)
Termitomyces clypeatus	edible (2)
Termitomyces entolomoides	edible (2)
Termitomyces eurhizus	edible (2)
Termitomyces heimii	edible (2)
Termitomyces microcarpus	edible (2)
Termitomyces striatus	edible (2)

Mauritius
1. Rammeloo and Walleyn, 1993; **2.** Walleyn and Rammeloo, 1994

Coprinus castaneus	edible (2)
Macrocybe spectabilis	[edible] (2)
Pseudohydnum gelatinosum	edible (1)
Tricholoma mauritianum	edible (1)
Volvariella volvacea	edible (1)

Mexico
1. Lopez, Cruz and Zamora-Martinez, 1992; **2.** Mata, 1987; **3.** Montoya-Esquivel, 1998; **4.** Montoya-Esquivel *et al.*, 2001; **5.** Moreno-Fuentes *et al.*, 1996; **6.** Richardson, 1991; **7.** Villarreal and Perez-Moreno, 1989; **8.** www.semarnat.gob.mx; **9.** Zamora-Martinez, Alvardo and Dominguez, 2000; **10.** Zamora-Martinez, Reygadas and Cifuentes, 1994

Agaricus arvensis	food (8)
Agaricus augustus	food (8)
Agaricus bisporus var. *albidus*	edible (7)
Agaricus bisporus var. *bisporus*	edible (7)
Agaricus bitorquis	food (8)
Agaricus campestris	food (8)
Agaricus comtulus	food (8)
Agaricus essettei	food (8)
Agaricus fuscofibrillosus	food (8)
Agaricus impudicus	food (8)
Agaricus placomyces	edible (8)
Agaricus silvaticus	food (8)
Agaricus silvicola	food (8)
A. squamuliferus var. *caroli*	food (8)
Agaricus subperonatus	food (8)
Agaricus subrutilescens	food (8)
Agrocybe vervacti	edible (10)
Albatrellus ovinus	food (8)
Aleuria aurantia	edible (7)
Amanita caesarea	food (8)
A. caesarea f. sp. *americana*	food (7)
Amanita calyptratoides	edible (7)
Amanita calyptroderma	edible (10)
Amanita ceciliae	food (8)
Amanita crocea	food (8)
Amanita flavivolva	[edible], medicinal, insecticidal (8)
Amanita flavoconia	food (8)
Amanita flavorubescens	edible (3)
Amanita fulva	food (8)
Amanita gemmata	edible (10)
Amanita inaurata	food (8)
Amanita muscaria	medicinal, insecticidal (8)
Amanita rubescens	food (8)
Amanita tuza	food (8)
Amanita umbonata	food (8)
Amanita vaginata	food (8)
Arachnion album	food (8)
Armillaria luteovirens	food (8)
Armillaria mellea	food (8)
Armillaria ostoyae	food (8)
Armillaria tabescens	food (8)
Auricularia auricula-judae	edible (8)

Auricularia delicata	edible (7)
Auricularia fuscosuccinea	edible (8)
Auricularia mesenterica	edible (8)
Auricularia polytricha	edible (8)
Boletellus ananas	food (8)
Boletellus betula	food (8)
Boletellus russellii	food (8)
Boletinus lakei	edible (7)
Boletus aestivalis	food (8)
Boletus atkinsonii	edible (3)
Boletus barrowsii	edible (7)
Boletus bicoloroides	food (8)
Boletus edulis	food (8)
Boletus erythropus	food (8)
Boletus felleus	edible (10)
Boletus frostii	food (8)
Boletus luridiformis	edible (3)
Boletus luridus	edible (7)
Boletus michoacanus	food (8)
Boletus pinicola	food (8)
Boletus pinophilus	food (4)
Boletus regius	edible (8)
Boletus reticulatus	food (8)
Boletus variipes	food (8)
Bovista plumbea var. *ovalispora*	food (8)
Chalciporus piperatus	edible (7)
Calvatia cyathiformis	food (8)
Camarophyllus pratensis	edible (7)
Cantharellula umbonata	edible (7)
Cantharellus cibarius	food (7)
Cantharellus odoratus	food (7)
Cantharellus tubiformis	food (8)
Chlorophyllum molybdites	edible (7)
Chroogomphus jamaicensis	food (4)
Chroogomphus rutilus	food (8)
Chroogomphus vinicolor	food (8)
Clavaria vermicularis	food (8)
Clavariadelphus cokeri	food (8)
Clavariadelphus pistillaris	food (8)
Clavariadelphus truncatus	food (8)
Clavariadelphus unicolor	food (8)
Clavicorona pyxidata	food (8)
Clavulina cinerea	food (8)
Clavulina cristata	edible (7)
Clavulina rugosa	edible (10)
Climacocystis borealis	edible (3)
Clitocybe clavipes	food (7)
Clitocybe gibba	food, medicinal (8)
Clitocybe nebularis	food (8)
Clitocybe odora	edible (8)
Clitocybe squamulosa	edible (3)
Clitocybe suaveolens	food (8)
Clitopilus prunulus	food (8)
Collybia acervata	edible (7)
Collybia butyracea	food (8)
Collybia confluens	food (8)
Collybia dryophila	food (4)
Collybia polyphylla	edible (8)
Cookeina sulcipes	edible (7)
Cookeina tricholoma	edible (7)
Coprinus comatus	edible (7)
Cortinarius glaucopus	food (4)
Craterellus cornucopioides	food (8)
Craterellus fallax	food (8)
Cronartium conigenum	edible (7)
Daldinia concentrica	medicinal (8)
Enteridium lycoperdon	edible (7)
Entoloma abortivum	food (7)
Entoloma clypeatum	food (4)
Favolus alveolarius	edible (7)
Favolus brasiliensis	edible (7)
Flammulina velutipes	food (8)
Fomitopsis pinicola	medicinal (8)
Fuligo septica	edible (7)
Ganoderma lobatum	medicinal (8)
Gautieria mexicana	edible (3)
Geastrum saccatum	medicinal (8)
Geastrum triplex	food (8); medicinal (2)
Gomphidius glutinosus	edible (7)
Gomphus clavatus	food (8)
Gomphus floccosus	food (8)
Gomphus kauffmanii	food (8)
Gyrodon merulioides	edible (7)
Gyromitra infula	food (8)
Gyroporus castaneus	edible (7)
Hebeloma fastibile	food (8)
Hebeloma mesophaeum	food (4)
Helvella acetabulum	food (4)
Helvella crispa	food (8)
Helvella elastica	food (8)
Helvella infula	food (4)
Helvella lacunosa	food (8)
Hericium caput-ursi	edible (7)
Hericium coralloides	edible (7)
Hericium erinaceus	food (8)
Hohenbuehelia petaloides	edible (7)
Hydnopolyporus fimbriatus	edible (7)
Hydnopolyporus palmatus	food (8)
Hydnum repandum	food (8)
Hygrocybe nigrescens	food (8)
Hygrophoropsis aurantiaca	food (8)
Hygrophorus chrysodon	food (8)
Hygrophorus niveus	food (8)
Hygrophorus purpurascens	food (8)
Hygrophorus russula	food (8)
Hypomyces lactifluorum	food (8)
Hypomyces macrosporus	edible (10)
Laccaria amethystina	food (8)
Laccaria bicolor	food (8)
Laccaria farinacea	edible (7)
Laccaria laccata	food (8)
Laccaria proxima	food (8)
Laccaria scrobiculatus	edible (1)
Lactarius carbonicola	edible (3)
Lactarius deliciosus	food (7)
Lactarius indigo	food (7)
Lactarius piperatus	food (8)
Lactarius salmonicolor	food (8)
Lactarius sanguifluus	edible (7)
Lactarius scrobiculatus	food (8)

Species	Use
Lactarius subdulcis	edible (10)
Lactarius vellereus	edible (7)
Lactarius volemus	food (8)
Lactarius yazooensis	food (4)
Laetiporus sulphureus	food (8)
Langermannia gigantea	food, medicinal (8)
Leccinum aurantiacum	food (8)
Leccinum chromapes	edible (7)
Leccinum rugosiceps	edible (3)
Lentinula boryana	food (7)
Lentinus conchatus	edible (7)
Lepiota aspera	edible (7)
Lepiota clypeolaria	edible (8)
Lepista irina	edible (7)
Lepista nuda	food (8)
Lepista personata	edible (7)
Lycoperdon candidum	edible (7)
Lycoperdon marginatum	edible (3)
Lycoperdon oblongisporum	edible (7)
Lycoperdon peckii	food (8)
Lycoperdon perlatum	food (7)
Lycoperdon pyriforme	food (8)
Lycoperdon rimulatum	edible (7)
Lycoperdon umbrinum	food (8)
L. umbrinum var. floccosum	edible (7)
Lyophyllum decastes	food (7)
Lyophyllum ovisporum	food (4)
Macrolepiota procera	edible (8)
Macropodia macropus	food (8)
Marasmius albogriseus	edible (7)
Marasmius oreades	food (8)
Melanoleuca evenosa	edible (7)
Melanoleuca grammopodia	edible (7)
Melanoleuca melaleuca	edible (7)
Merulius incarnatus	food (8)
Morchella angusticeps	edible (10)
Morchella conica	food (8)
Morchella costata	edible (7)
Morchella crassipes	food (8)
Morchella elata	food (8)
Morchella esculenta	food (8)
Mycena pura	food (8)
Neolentinus lepideus	edible (8)
Neolentinus ponderosus	food (5)
Oudemansiella canarii	food (8)
Panus crinitus	edible (7)
Paxina acetabulum	food (8)
Peziza badia	food (8)
Pholiota lenta	food (4)
Pleurotus cornucopiae	edible (7)
Pleurotus djamor	food (8)
Pleurotus dryinus	food (8)
Pleurotus levis	food (8)
Pleurotus ostreatoroseus	edible (7)
Pleurotus ostreatus	food (7); medicinal (8)
Pleurotus smithii	edible (7)
Pluteus aurantiorugosus	food (8)
Pluteus cervinus	food (7)
Pogonomyces hydnoides	food (8)
Psathyrella spadicea	edible (10)
Pseudohydnum gelatinosum	edible (8)
Psilocybe zapotecorum	edible, hallucinogen (8)
Pycnoporus sanguineus	medicinal (8)
Ramalina ecklonii	edible (8)
Ramaria aurea	food (7)
Ramaria bonii	edible (3)
Ramaria botrytis	food (8)
Ramaria botrytoides	edible (3)
Ramaria cystidiophora	edible (3)
Ramaria flava	edible (8)
Ramaria flavobrunnescens	food (7)
Ramaria rosella	edible (3)
Ramaria rubiginosa	food (8)
Ramaria rubripermanens	food (4)
Ramaria sanguinea	edible (3)
Ramaria stricta	edible (7)
Rhizopogon	food (8)
Rhodophyllus clypeatus	food (8)
Roccella	other – dye (6)
Rozites caperatus	food (8)
Russula aciculocystis	edible (3)
Russula alutacea	food (8)
Russula brevipes	food (7)
Russula cyanoxantha	food (8)
Russula delica	food (4)
Russula densifolia	edible (7)
Russula emetica	edible (9)
Russula lepida	food (8)
Russula lutea	food (8)
Russula macropoda	edible (3)
Russula mariae	food (4)
Russula mexicana	edible (10)
Russula nigricans	food (8)
Russula olivacea	food (8)
Russula ornaticeps	edible (3)
Russula queletii	edible (10)
Russula romagnesiana	food (4)
Russula rubroalba	edible (3)
Russula vesca	edible (7)
Russula xerampelina	food (4)
Sarcodon imbricatus	food (8)
Sarcoscypha coccinea	food (8)
Sarcosphaera eximia	food (4)
Schizophyllum commune	edible (7)
Schizophyllum fasciatum	edible (7)
Sparassis crispa	food (8)
Strobilomyces confusus	edible (7)
Strobilomyces floccopus	food (8)
Stropharia coronilla	food (4)
Suillus acidus	edible (7)
Suillus americanus	food (8)
Suillus brevipes	food (8)
Suillus cavipes	food (8)
Suillus granulatus	food (8)
Suillus hirtellus	food (8)
Suillus luteus	food (8)
Suillus pseudobrevipes	food (4)
Suillus tomentosus	food (8)
Tephrocybe atrata	edible (10)
Thelephora paraguayensis	medicinal (2)

Trametes versicolor	medicinal (8)
Tremella concrescens	edible (8)
Tremellodendron schweinitzii	edible (8)
Tricholoma flavovirens	food (8)
Tricholoma magnivelare	food (8)
Tricholoma sejunctum	food (8)
Tricholoma ustaloides	edible (10)
Tricholoma vaccinum	edible (10)
Tylopilus felleus	food (4)
Ustilago maydis	food (7)
Vascellum curtisii	edible (7)
Vascellum intermedium	food (8)
Vascellum pratense	edible (7), medicinal (8)
Vascellum qudenii	food, medicinal (8)
Volvariella bombycina	edible (7)
Volvariella volvacea	edible (7)
Xanthoconium separans	edible (7)
Xerocomus badius	edible (7)
Xerocomus chrysenteron	edible (8)
Xerocomus spadiceus	edible (8)
Xeromphalina campanella	medicinal (8)

MOZAMBIQUE
1. Uaciquete, Dai and Motta, 1996; 2. Wilson, Cammack and Shumba, 1989

Afroboletus luteolus	food (1)
Amanita hemibapha	food (2)
Armillaria mellea	food (1)
Auricularia auricula-judae	food (2)
Boletus edulis	food (1)
Cantharellus cibarius	food (2)
Cantharellus densifolius	food (1)
Cantharellus longisporus	food (2)
Cantharellus pseudocibarius	food (1)
Cantharellus symoensii	food (1)
Coprinus micaceus	food (1)
Lentinus squarrulosus	food (2)
Leucoagaricus leucothites	food (1)
Micropsalliota brunneosperma	food (2)
Phlebopus colossus	food (2)
Psathyrella candolleana	food (2)
Schizophyllum commune	food (1)
Termitomyces	food (1)
Termitomyces clypeatus	food (2)
Termitomyces eurhizus	food (2)
Termitomyces microcarpus	food (2)
Termitomyces schimperi	food (2)

MOROCCO
1. Alsheikh and Trappe, 1983; 2. Kytovuori, 1989; 3. Moreno-Arroyo *et al.*, 2001; 4. Richardson, 1991; 5. FAO, 2001b

Agaricus bisporus	edible (5)
Boletus edulis	edible (5)
Cantharellus cibarius	edible (5)
Evernia prunastri	other – perfume (4)
Morchella sp.	edible (5)
Pleurotus ostreatus	edible (5)
Pseudevernia furfuracea	other – perfume (4)
Terfezia leonis	edible (5)
Tirmania nivea	edible (1)
Tricholoma caligatum	edible (5)
Tricholoma nauseosum	edible (2)
Tuber oligospermum	edible (3)

MYANMAR
Pegler and Vanhaecke, 1994

Termitomyces eurhizus	edible

NAMIBIA
1. Rammeloo and Walleyn, 1993; 2. Taylor *et al.*, 1995; 3. Walleyn and Rammeloo, 1994

Battarrea stevenii	medicinal; cosmetic (3)
Terfezia pfeilii	food (2)
Termitomyces schimperi	edible (1)
Termitomyces umkowaanii	edible (1)

NEPAL
1. Adhikari, 1999; 2. Adhikari and Durrieu, 1996; 3. Richardson, 1991; 4. Zang and Doi, 1995

Agaricus bitorquis	food (1)
Agaricus campestris	edible (1)
Agaricus silvicola	food (1)
Agaricus subrufescens	food (1)
Amanita caesarea	food (1)
Amanita chepangiana	edible (2)
Amanita hemibapha	edible (1)
Amanita vaginata	edible (2)
Armillaria mellea	edible (1)
Astraeus sp.	edible (2)
Auricularia auricula-judae	edible (1)
Auricularia delicata	edible (1)
Auricularia mesenterica	edible (1)
Auricularia polytricha	edible (1)
Boletus edulis	edible (2)
Boletus luridus	edible (2)
Boletus vitellinus	edible (2)
Cantharellus cibarius	food (1)
Cantharellus odoratus	edible (1)
Cantharellus subalbidus	edible (2)
Cantharellus subcibarius	edible (1)
Cantharellus tubiformis	edible (2)
Clavaria vermicularis	edible (2)
Clavulina cinerea	food (1)
Clavulina cristata	food (1)
Clavulinopsis fusiformis	edible (2)
Collybia butyracea	edible (2)
Coprinus comatus	edible (2)
Cordyceps sinensis	medicinal (1)
Craterellus cornucopioides	edible (1)
Crepidotus mollis	[edible] (2)
Evernia prunastri	other – perfume (96)
Fibroporia vaillantii	medicinal (2)
Fistulina hepatica	medicinal (2)
Flammulina velutipes	edible (1)
Ganoderma applanatum	medicinal (2)
Ganoderma lucidum	[medicinal] (2)

Geastrum sp.	edible (2)
Grifola frondosa	food (1)
Hericium clathroides	edible (1)
Hericium coralloides	food (1)
Hericium erinaceus	food (1)
Hericium flagellum	food (1)
Hericium laciniatum	edible (2)
Hydnum ranceo-foetidum	[edible] (1)
Hydnum repandum	food (1)
Inonotus hispidus	medicinal (2)
Laccaria amethystina	food (1)
Laccaria laccata	food (1)
Lactarius deliciosus	food (2)
Lactarius piperatus	edible (2)
Lactarius volemus	edible (2)
Laetiporus sulphureus	food (1)
Lentinula edodes	food (1)
Lycoperdon sp.	edible (2)
Macrolepiota procera	edible (2)
Marasmius oreades	edible (2)
Meripilus giganteus	food (2)
Morchella conica	edible (1)
Morchella deliciosa	edible (1)
Morchella elata	[edible] (1)
Morchella esculenta	edible (1)
Morchella smithiana	[edible] (1)
Morchella esculenta var. *vulgaris*	edible (1)
Pholiota nameko	edible (2)
Pleurotus circinatus	edible (1)
Pleurotus cornucopiae	edible (1)
Pleurotus dryinus	food (1)
Pleurotus nepalensis	edible (1)
Pleurotus ostreatus	food (1)
P. ostreatus var. *magnificus*	edible (1)
Pleurotus pulmonarius	edible (1)
Plüteus cervinus	food (1)
Polyporus arcularius	food (1)
Polyporus badius	edible (1)
Polyporus brumalis	medicinal (2)
Pycnoporus cinnabarinus	edible (2)
Ramaria aurea	food (2)
Ramaria botrytis	food (1)
Ramaria flava	food (2)
Ramaria formosa	edible (2)
Ramaria fuscobrunnea	food (1)
Ramaria obtusissima	food (1)
Rhizopogon luteolus	edible (2)
Russula chloroides	food (2)
Russula delica	edible (2)
Russula nigricans	edible (2)
Russula virescens	food (2)
Scleroderma citrinum	edible (1)
Scleroderma texense	edible (1)
Secotium himalaicum	edible (149)
Termitomyces eurhizus	food (1)
Trametes hirsuta	medicinal (2)
Tremella mesenterica	edible (2)
Volvariella volvacea	food (1)
Xerula radicata	food (1)

NIGERIA

1. Alofe, Odeyemi and Oke, 1996; **2.** Oso, 1975; **3.** Rammeloo and Walleyn, 1993; **4.** Walleyn and Rammeloo, 1994

Agrocybe broadwayi	food (2)
Armillaria mellea	edible (3)
Auricularia auricula-judae	food (2)
Calvatia cyathiformis	food, medicinal (2)
Chlorophyllum molybdites	edible (3)
Coprinus africanus	food (2)
Lentinus subnudus	edible (1)
Lentinus tuber-regium	food (2); medicinal, cosmetic (4)
Lentinus velutinus	medicinal (4)
Macrocybe lobayensis	food (2)
Panus flavus	medicinal (2)
Phallus aurantiacus	[poisonous], medicinal (4)
Pleurotus squarrosulus	food (2)
Psathyrella atroumbonata	food (2)
Schizophyllum commune	food (2)
Termitomyces clypeatus	food (2)
Termitomyces globulus	food (2); animal poison (4)
Termitomyces mammiformis	food (2)
Termitomyces microcarpus	food (2); medicinal (4)
Termitomyces robustus	food (2)
Termitomyces striatus	edible (3)
Volvariella esculenta	food (2)
Volvariella volvacea	food (2)

PAKISTAN

1. Batra, 1983; **2.** Gardezi, 1993; **3.** FAO, 1993b; **4.** Pegler and Vanhaecke, 1994; **5.** Syed-Riaz and Mahmood-Khan, 1999

Agaricus augustus	edible (2)
Agaricus campestris	edible (2)
Agaricus placomyces	edible (2)
Agaricus rodmani	edible (2)
Agaricus silvaticus	edible (2)
Agaricus silvicola	edible (2)
Armillaria mellea	edible (5)
Cantharellus cibarius	edible (5)
Craterellus cornucopioides	edible (5)
Flammulina velutipes	edible (5)
Macrolepiota procera	edible (5)
Morchella angusticeps	edible (3)
Morchella conica	edible (3)
Morchella esculenta	edible (3)
Podaxis pistillaris	edible (1)
Termitomyces clypeatus	edible (4)
Termitomyces eurhizus	edible (4)
Termitomyces heimii	edible (4)
Termitomyces microcarpus	edible (4)
Termitomyces radicatus	edible (4)
Termitomyces striatus	edible (4)

Papua New Guinea Sillitoe, 1995	
Armillaria sp.	not eaten
Auricularia polytricha	not eaten
Boletus erythropus var. *novoguineensis*	edible
Boletus nigroviolaceus	edible
Bondarzewia montana	edible
Cantharellus	edible
Collybia sp.	not eaten
Cortinarius sp.	edible
Grifola frondosa	edible
Gymnopilus novoguineensis	not eaten
Inocybe sp.	edible
Laccaria amethystea	edible
Lactarius	edible
Lentinula lateritia	edible
Lentinus araucariae	edible
Lentinus umbrinus	not eaten
Microporus affinis	edible
Microporus xanthopus	not eaten
Oudemansiella canarii	edible
Phaeomarasmius affinis	edible
Phellinus senex	not eaten
Pholiota austrospumosa	edible
Phylloporus bellus	not eaten
Pleurotus djamor	edible
Polyporus arcularius	edible
Polyporus blanchetianus	edible
Polyporus tenuiculus	edible
Pycnoporus coccineus	other – raw material
Pycnoporus sanguineus	edible
Ramaria fistulosa	edible
Russula amaendum	edible
Russula eburneoareolata	edible
Russula pseudoamaendum	edible
Strobilomyces velutipes	edible
Trametes versicolor	not eaten
Trogia sp.	edible

Peru **1.** Diez, 2003, personal communication: *Collecting* Boletus edulis *for commercial purposes in Peru*; **2.** Remotti and Colan, 1990	
Auricularia delicata	edible (2)
Auricularia fuscosuccinea	edible (2)
Boletus edulis	food (1)
Favolus alveolarius	edible (2)
Favolus brasiliensis	edible (2)
Lentinus conchatus	edible (2)
Pleurotus concavus	edible (2)
Pleurotus ostreatus	edible (2)
Pleurotus roseopileatus	edible (2)
Pluteus cervinus	edible (2)
Polyporus arcularius	edible (2)
Polyporus sanguineus	edible (2)
Schizophyllum brevilamellatum	edible (2)
Schizophyllum commune	edible (2)
Volvariella bakeri	edible (2)

Philippines **1.** Novellino, 1999; **2.** Pegler and Vanhaecke, 1994. See also Mendoza, 1938 – records not included	
Agaricus ?spp.	food (1)
Ganoderma ?spp.	food (1)
Pleurotus ?spp.	food (1)
Polyporus ?spp.	food (1)
Termitomyces eurhizus	edible (2)
Termitomyces microcarpus	edible (2)
Termitomyces striatus	edible (2)

Poland www.grzyby.pl	
Armillaria meilea	food
Auricularia auricula-judae	food
Boletus edulis	food
Cantharellus cibarius	food
Lactarius deliciosus	food
Leccinum griseum	food
Leccinum scabrum	food
Macrolepiota procera	food
Pleurotus ostreatus	food
Rozites caperatus	food
Russula cyanoxantha	food
Tricholoma equestre	food
Xerocomus badius	food
Xerocomus subtomentosus	food

Réunion Rammeloo and Walleyn, 1993	
Volvariella volvacea	edible

Russian Federation **1.** Saar, 1991; **2.** Vasil'eva, 1978. Note: This is only for the Russian far east.	
Agaricus campestris	edible (2)
Agaricus placomyces	edible (2)
Agaricus silvaticus	edible (2)
Agaricus silvicola	edible (2)
Aleuria aurantia	[edible] (2)
Amanita caesareoides	edible (2)
Amanita crocea	edible (2)
Amanita muscaria	poisonous (2); medicinal (1)
Amanita vaginata	edible (2)
Armillaria mellea	edible (2)
Auricularia auricula-judae	edible (2)
Auricularia polytricha	edible (2)
Boletinus asiaticus	edible (2)
Boletinus paluster	not known (2)
Boletus calopus	edible (2)
Boletus edulis	not edible (2)
Boletus erythropus	edible (2)
Boletus luridus	edible (2)
Boletus regius	edible (2)
B. tomentososquamulosus	not edible (2)
Bovista plumbea	edible (2)
Buchwaldoboletus spectabilis	edible (2)
Calocybe gambosa	edible (2)

Calvatia excipuliformis	edible (2)
Calvatia utriformis	edible (2)
Camarophyllus niveus	edible (2)
Camarophyllus pratensis	edible (2)
Camarophyllus virgineus	not known (2)
Cantharellus cibarius	edible (2)
Cantharellus floccosus	edible (2)
Catathelasma ventricosum	edible (2)
Chalciporus piperatus	edible (2)
Chroogomphus rutilus	edible (2)
Clavaria purpurea	edible (2)
Clavariadelphus pistillaris	edible (2)
Clavariadelphus sachalinensis	edible (2)
Clavariadelphus truncatus	edible (2)
Clavulina amethystina	edible (2)
Clavulina cristata	edible (2)
Clitocybe infundibuliformis	edible (2)
Clitocybe nebularis	edible (2)
Clitocybe odora	edible (2)
Clitocybe suaveolens	edible (2)
Clitopilus prunulus	edible (2)
Collybia contorta	edible (2)
Collybia dryophila	edible (2)
Coprinus atramentarius	edible (2)
Coprinus comatus	edible (2)
Coprinus micaceus	edible (2)
Cortinarius alboviolaceus	edible (2)
Cortinarius armeniacus	edible (2)
Cortinarius armillatus	edible (2)
Cortinarius collinitus	edible (2)
Cortinarius glaucopus	edible (2)
Cortinarius orichalceus	edible (2)
Cortinarius prasinus	edible (2)
Craterellus cornucopioides	edible (2)
Flammulina velutipes	edible (2)
Fomes fomentarius	medicinal (1)
Gomphidius maculatus	edible (2)
Gomphidius purpurascens	edible (2)
Gomphus clavatus	edible (2)
Gyromitra ambigua	edible (2)
Gyromitra esculenta	not known (2)
Gyromitra infula	not known (2)
Gyromitra ussuriensis	edible (2)
Helvella crispa	edible (2)
Hericium erinaceus	edible (2)
Hydnotrya tulasnei	edible (2)
Hydnum repandum	edible (2)
Hygrocybe cantharellus	edible (2)
Hygrocybe coccinea	edible (2)
Hygrocybe conica	edible (2)
Hygrocybe laeta	edible (2)
Hygrocybe obrussea	edible (2)
Hygrocybe psittacina	edible (2)
Hygrocybe punicea	edible (2)
Hygrocybe unguinosa	edible (2)
Hygrophorus agathosmus	edible (2)
Hygrophorus camarophyllus	edible (2)
Hygrophorus chrysodon	edible (2)
Hygrophorus eburneus	edible (2)
Hygrophorus erubescens	edible (2)
Hygrophorus limacinus	edible (2)
Hygrophorus lucorum	edible (2)
Hygrophorus olivaceoalbus	edible (2)
Hygrophorus pudorinus	edible (2)
Hygrophorus russula	edible (2)
Inonotus obliquus	medicinal (1)
Kuehneromyces mutabilis	edible (2)
Laccaria amethystina	edible (2)
Laccaria laccata	edible (2)
Lactarius chrysorrheus	edible (2)
Lactarius controversus	edible (2)
Lactarius deliciosus	edible (2)
Lactarius flavidulus	edible (2)
Lactarius insulsus	edible (2)
Lactarius japonicus	edible (2)
Lactarius necator	edible (2)
Lactarius piperatus	edible (2)
Lactarius pubescens	edible (2)
Lactarius pyrogalus	edible (2)
Lactarius repraesentaneus	[edible] (2)
Lactarius resimus	edible (2)
Lactarius rufus	edible (2)
Lactarius scrobiculatus	edible (2)
Lactarius torminosus	edible (2)
Lactarius trivialis	edible (2)
Lactarius uvidus	[edible] (2)
Lactarius vellereus	edible (2)
Lactarius volemus	edible (2)
Laetiporus sulphureus	edible (2)
Langermannia gigantea	edible (2)
Leccinum aurantiacum	edible (2)
Leccinum chromapes	edible (2)
Leccinum extremiorientale	edible (2)
Leccinum holopus	not known (2)
Leccinum oxydabile	edible (2)
Leccinum scabrum	edible (2)
Leccinum testaceoscabrum	edible (2)
Lepista glaucocana	edible (2)
Leucoagaricus leucothites	edible (2)
Leucocortinarius bulbiger	edible (2)
Limacella illinita	edible (2)
Lycoperdon perlatum	edible (2)
Lycoperdon pyriforme	[edible] (2)
Lyophyllum connatum	edible (2)
Lyophyllum decastes	edible (2)
Lyophyllum ulmarium	edible (2)
Macrolepiota procera	edible (2)
Macrolepiota puellaris	edible (2)
Marasmius oreades	edible (2)
Marasmius scorodonius	edible (2)
Melanoleuca brevipes	edible (2)
Melanoleuca grammopodia	edible (2)
Melanoleuca verrucipes	not known (2)
Morchella conica	edible (2)
Morchella esculenta	edible (2)
Otidea onotica	edible (2)
Oudemansiella brunneomarginata	edible (2)
Oudemansiella mucida	edible (2)
Panellus serotinus	edible (2)
Paxillus involutus	edible (2)
Phaeolepiota aurea	edible (2)

Phallus impudicus	not edible (2)
Phellinus igniarius	medicinal (1)
Pholiota aurivella	edible (2)
Pleurotus citrinopileatus	edible (2)
Pleurotus ostreatus	edible (2)
Plicaria badia	edible (2)
Pluteus cervinus	edible (2)
Pluteus coccineus	edible (2)
Polyporus squamosus	edible (2)
Porphyrellus atrobrunneus	edible (2)
Porphyrellus pseudoscaber	edible (2)
Pseudohydnum gelatinosum	edible (2)
Psiloboletinus lariceti	edible (2)
Ptychoverpa bohemica	edible (2)
Ramaria aurea	edible (2)
Ramaria botrytoides	edible (2)
Ramaria flava	edible (2)
Ramaria formosa	not edible (2)
Ramaria invalii	not edible (2)
Ramaria obtusissima	not edible (2)
Ramaria pulcherrima	edible (2)
Rhizopogon roseolus	edible (2)
Rhodophyllus aprilis	edible (2)
Rhodophyllus clypeatus	edible (2)
Rozites caperatus	edible (2)
Russula adusta	edible (2)
Russula aeruginea	edible (2)
Russula albonigra	edible (2)
Russula alutacea	edible (2)
Russula aurata	edible (2)
Russula consobrina	edible (2)
Russula cyanoxantha	edible (2)
Russula delica	edible (2)
Russula emetica	edible (2)
Russula flava	edible (2)
Russula foetens	edible (2)
Russula fragilis	edible (2)
Russula olivascens	edible (2)
Russula pectinatoides	edible (2)
Russula punctata	edible (2)
Russula queletii	not known (2)
Russula vesca	edible (2)
Russula virescens	edible (2)
Russula xerampelina	edible (2)
Sarcodon imbricatus	edible (2)
Sarcodon lobatus	edible (2)
Sarcoscypha coccinea	edible (2)
Scutiger ovinus	edible (2)
Sparassis crispa	edible (2)
Strobilomyces floccopus	edible (2)
Stropharia rugosoannulata	edible (2)
Suillus abietinus	edible (2)
Suillus americanus	edible (2)
Suillus bovinus	edible (2)
Suillus cavipes	edible (2)
Suillus granulatus	edible (2)
Suillus grevillei	edible (2)
Suillus luteus	edible (2)
Suillus pictus	edible (2)
Suillus placidus	edible (2)
Suillus plorans	edible (2)
Suillus subluteus	edible (2)
Suillus variegatus	edible (2)
Suillus viscidus	edible (2)
Tremiscus helvelloides	edible (2)
Tricholoma atrosquamosum	edible (2)
Tricholoma fulvum	edible (2)
Tricholoma orirubens	edible (2)
Tricholoma portentosum	edible (2)
Tricholoma terreum	edible (2)
Tricholomopsis decora	edible (2)
Tricholomopsis rutilans	edible (2)
Tylopilus neofelleus	not edible (2)
Volvariella speciosa	edible (2)
Xerocomus badius	edible (2)
Xerocomus chrysenteron	edible (2)
Xerocomus rubellus	edible (2)
Xerocomus subtomentosus	edible (2)

Saudi Arabia
1. Alsheikh and Trappe, 1983; **2.** Bokhary and Parvez, 1993; **3.** Kirk *et al*., 2001

Parmelia austrosinensis	food (3)
Terfezia claveryi	edible (2)
Tirmania nivea	edible (1)

Senegal
1. Ducousso, Ba and Thoen, 2002; **2.** Thoen and Ba, 1989

Afroboletus costatisporus	[edible] (2)
Amanita crassiconus	[edible] (2)
Amanita hemibapha	[edible] (2)
Amanita rubescens	[edible] (2)
Cantharellus congolensis	[edible] (2)
Cantharellus pseudofriesii	[edible] (2)
Gyrodon intermedius	food (1)
Lactarius gymnocarpus	[edible] (2)
Phlebopus sudanicus	food (1)
Polyporus	medicinal (122)
Russula foetens	[edible] (2)
Russula pectinata	[edible] (2)
Tubosaeta brunneosetosa	[edible] (2)

Sierra Leone
Pegler and Vanhaecke, 1994

Termitomyces striatus	edible

Singapore
Burkhill, 1935

Termitomyces albuminosus	food

Slovenia
www.matkurja.com

Agaricus bitorquis	edible
Agaricus campestris	edible
Amanita caesarea	edible
Amanita rubescens	edible
Armillaria mellea	edible

Astraeus hygrometricus	not edible
Boletus aestivalis	edible
Boletus erythropus	edible
Calocybe gambosa	edible
Cantharellus cibarius	edible
Coprinus comatus	edible
Craterellus cornucopioides	edible
Leccinum griseum	edible
Leccinum scabrum	edible
Leccinum testaceoscabrum	edible
Macrolepiota procera	edible
Macrolepiota rhacodes	edible
Morchella esculenta	edible
Pleurotus ostreatus	edible
Russula cyanoxantha	edible
Tricholoma portentosum	edible
Xerocomus badius	edible
Xerocomus subtomentosus	edible

Somalia
Rammeloo and Walleyn, 1993

Agaricus amboensis	edible
Agaricus campestris	edible

South Africa
1. Pegler and Vanhaecke, 1994; **2.** Walleyn and Rammeloo, 1994

Amanita excelsa	[edible] (2)
Amanita foetidissima	[edible] (2)
Amanita muscaria	hallucinogen, poisonous (2)
Amanita rubescens	[edible] (2)
Helvella lacunosa	[edible] (2)
Hericium coralloides	[edible] (2)
Lepista caffrorum	[edible] (2)
Macrolepiota rhacodes	[edible] (2)
Psilocybe semilanceata	hallucinogen (2)
Suillus granulatus	[edible] (2)
Termitomyces striatus	edible (1)

Spain
1. Cervera and Colinas 1997; **2.** Martinez, Oria de Rueda and Martinez, 1997; **3.** Martinez, Florit and Colinas (1997)

Agaricus arvensis	food (2)
Agrocybe aegerita	food (2)
Amanita caesarea	food (2)
Amanita ponderosa	food (2)
Armillaria mellea	food (2)
Boletus aereus	food (2)
Boletus aestivalus	food (2)
Boletus edulis	food (2)
Boletus pinicola	food (2)
Boletus regius	food (2)
Boletus reticulatus	food (2)
Calocybe gambosa	food (2)
Cantharellus cibarius	food (2)
Cantharellus lutescens	food (2)
Cantharellus tubaeformis	food (2)
Clitocybe geotropa	food (2)
Clitocybe nebularis	food (2)
Coprinus comatus	food (2)
Craterellus cornucopioides	food (2)
Helvella leucomelaena	food (2)
Helvella monachella	food (2)
Hydnum repandum	food (2)
Hydnum rufescens	food (2)
Hygrophorus eburneus	food (1)
Hygrophorus latitabundus	food (3)
Hygrophorus limacinus	food (2)
Hygrophorus olivaceoalbus	food (2)
Hygrophorus russula	food (1)
Lactarius deliciosus	food (2)
Lactarius sanguifluus	food (2)
Leccinum aurantiacum	food(2)
Leccinum lepidum	food (2)
Lepista nuda	food (2)
Lepista personata	food (2)
Leucopaxillus candidus	food (2)
Leucopaxillus lepistoides	food (2)
Macrolepiota procera	food (2)
Macrolepiota rhacodes	food (2)
Marasmius oreades	food (2)
Morchella esculenta	food (2)
Pleurotus eryngii	food (2)
Pleurotus nebrodensis	food (2)
Pleurotus ostreatus	food (2)
Rhodocybe truncata	food (2)
Russula cyanoxantha	food(2)
Russula virescens	food (2)
Suillus bellinii	food (2)
Suillus bovinus	food (3)
Suillus granulatus	food (2)
Suillus luteus	food(2)
Suillus variegatus	food (3)
Terfezia arenaria	food (2)
Terfezia claveryi	food (2)
Terfezia leptoderma	food (2)
Tricholoma equestre	food (2)
Tricholoma goniospermum	food (2)
Tricholoma portentosum	food (2)
Tricholoma terreum	food (2)
Tuber aestivum	food (2)
Tuber brumale	food (2)
Tuber melanosporum	food (2)

Sri Lanka
Pegler and Vanhaecke, 1994

Termitomyces eurhizus	edible
Termitomyces microcarpus	edible

Tanzania [United Republic of]
1. Härkönen, Saarimäki and Mwasumbi, 1994a; **2.** Härkönen, Saarimäki and Mwasumbi, 1994b; **3.** Rammeloo and Walleyn, 1993; **4.** Walleyn and Rammeloo, 1994

Agaricus campestris	edible (3)
Amanita tanzanica	edible (2)
Amanita zambiana	edible (2)

Armillaria mellea	edible (2)
Auricularia delicata	edible (2)
Auricularia fuscosuccinea	edible (2)
Auricularia polytricha	edible (2)
Cantharellus congolensis	edible (2)
Cantharellus isabellinus	edible (2)
Cantharellus platyphyllus	edible (2)
Cantharellus symoensii	edible (2)
Coprinus cinereus	edible (2)
Entoloma argyropus	edible (3)
Hypholoma subviride	not eaten (2)
Kuehneromyces mutabilis	edible (3)
Lactarius gymnocarpus	edible (2)
Lactarius kabansus	food (2)
Lactarius pelliculatus	edible (2)
Lactarius phlebophyllus	food (2)
Lactarius rubroviolascens	edible (2)
Lentinus sajor-caju	edible (3)
Lentinus tuber-regium	edible (3), medicinal (4)
Lenzites elegans	edible (3)
Leucoagaricus leucothites	edible (3)
Leucoagaricus rhodocephalus	edible (4)
Lignosus sacer	medicinal (4)
Macrolepiota procera	edible (3)
Phellinus sp.	medicinal (4)
Pleurotus djamor	edible (2)
Polyporus moluccensis	edible (2)
Russula cellulata	food (2)
Russula ciliata	edible (2)
Russula compressa	edible (2)
Russula congoana	edible (2)
Russula heimii	edible (1)
Russula hiemisilvae	edible (2)
Russula liberiensis	edible (1)
Russula phaeocephala	edible (1)
Russula sublaevis	edible (1)
Russula tanzaniae	edible (1)
Suillus granulatus	edible (2)
Termitomyces aurantiacus	edible (2)
Termitomyces eurhizus	edible, medicinal (2)
Termitomyces letestui	food (2)
Termitomyces microcarpus	edible (2)
Termitomyces singidensis	food (2)
Volvariella bombycina	edible (3)
Volvariella volvacea	edible (3)

THAILAND
1. Jones, Whalley and Hywel-Jones, 1994; **2.** Pegler and Vanhaecke, 1994; **3.** Stamets, 2000

Auricularia sp.	food (1)
Cantharellus cibarius	food (1)
Cantharellus minor	food (1)
Lentinula edodes	food (1)
Lentinus praerigidus	food (1)
Pleurotus cystidiosus	food (3)
Russula aeruginea	food (1)
Russula delica	food (1)
Russula densifolia	food (1)
Russula foetens	food (1)
Russula heterophylla	food (1)
Russula lepida	food (1)
Russula nigricans	food (1)
Russula sanguinea	food (1)
Russula violeipes	food (1)
Russula virescens	food (1)
Termitomyces aurantiacus	food (2)
Termitomyces clypeatus	food (2)
Termitomyces globulus	food (2)
Volvariella volvacea	food (1)

TUNISIA
Alsheikh and Trappe, 1983

Tirmania nivea	edible

TURKEY
1. Afyon, 1997; **2.** Caglarirmak, Unal and Otles, 2002; **3.** Demirbas, 2000; **4.** Sabra and Walter, 2001; **5.** http//www.ogm.gov.tr/; **6.** Yilmaz, Oder and Isiloglu, 1997

Agaricus bisporus	food (6)
Agaricus bitorquis	edible (3)
Agaricus campestris	food (6)
Agaricus silvicola	edible (3)
Amanita caesarea	edible (5)
Armillaria mellea	edible (5)
Boletus edulis	food (4)
Cantharellus cibarius	food (4)
Chroogomphus rutilus	edible (5)
Coprinus comatus	food (1)
Cortinarius variecolor	edible (5)
Craterellus cornucopioides	edible (5)
Fistulina hepatica	edible (5)
Helvella lacunosa	food (1)
Hericium coralloides	food (6)
Hydnum repandum	edible (5)
Hygrophorus chrysodon	edible (5)
Laccaria laccata	edible (3)
Lactarius deliciosus	food (6)
Lactarius piperatus	food (2)
Lactarius salmonicolor	food (6)
Lactarius volemus	edible (5)
Laetiporus sulphureus	edible (5)
Lycoperdon perlatum	food (6)
Macrolepiota procera	edible (5)
Morchella conica	food (6)
Morchella crassipes	edible (1)
Morchella deliciosa	edible (5)
Morchella elata	edible (1)
Morchella esculenta	food (6)
M. esculenta var. *rotunda*	edible (5)
Pleurotus cornucopiae	edible (5)
Pleurotus eryngii	food (1)
Pleurotus ostreatus	food (6)
Polyporus squamosus	edible (5)
Rhizopogon luteolus	edible (5)
Rhizopogon roseolus	food (6)
Rhizopogon rubescens	edible (5)
Russula delica	food (6)
Sparassis crispa	edible (5)

Suillus bovinus	edible (5)
Suillus grevillei	edible (5)
Suillus luteus	edible (5)
Terfezia boudieri	food (4)
Tricholoma populinum	food (1)
Tricholoma terreum	edible (5)
Tuber aestivum	edible (4)
Tuber borchii	edible (4)
Xerocomus badius	edible (5)

UGANDA
1. Katende, Segawa and Birnie, 1999; **2.** Pegler and Vanhaecke, 1994; **3.** Rammeloo and Walleyn, 1993

Agaricus bingensis	edible (3)
Armillaria mellea	edible (1)
Lentinus prolifer	edible (1)
Termitomyces aurantiacus	edible (1)
Termitomyces eurhizus	edible (1)
Termitomyces letestui	edible (1)
Termitomyces microcarpus	edible (1)
Termitomyces robustus	edible (3)
Termitomyces striatus	edible (2)
Tricholoma sp.	edible (3)

UKRAINE
Zerova and Rozhenko, 1988

Agaricus arvensis	[edible]
Agaricus bisporus	[edible]
Agaricus bitorquis	[edible]
Agaricus campestris	[edible]
Agaricus macrosporus	[edible]
Agaricus placomyces	[edible]
Agaricus silvaticus	[edible]
Amanita caesarea	[edible]
Amanita excelsa	[edible]
Amanita porphyria	[edible]
Amanita rubescens	[edible]
Amanita vaginata	[edible]
Amanita xanthodermus	[edible]
Armillaria mellea	[edible]
Astrosporina asterospora	[edible]
Boletus appendiculatus	[edible]
Boletus aurantiacus	[edible]
Boletus calopus	[edible]
Boletus edulis	[edible]
Boletus elegans	[edible]
Boletus erythropus	[edible]
Boletus impolitus	[edible]
Boletus luridus	[edible]
Boletus regius	[edible]
Boletus rubellus	[edible]
Boletus scaber	[edible]
Boletus subtomentosus	[edible]
Boletus variegatus	[edible]
Calvatia utriformis	[edible]
Cantharellus cibarius	[edible]
Chalciporus piperatus	[edible]
Clitocybe aurantiaca	[edible]
Clitocybe clavipes	[edible]
Clitocybe nebularis	[edible]
Clitocybe olearia	[edible]
Clitocybe rivulosa	[edible]
Clitopilus prunulus	[edible]
Collybia butyracea	[edible]
Coprinus comatus	[edible]
Coprinus micaceus	[edible]
Cortinarius crassus	[edible]
Cortinarius mucosus	[edible]
Cortinarius multiformis	[edible]
Cortinarius varius	[edible]
Entoloma clypeatum	[edible]
Entoloma rhodopolium	[edible]
Flammulina velutipes	[edible]
Gomphidius glutinosus	[edible]
Gyrodon lividus	[edible]
Gyromitra esculenta	[edible]
Gyroporus castaneus	[edible]
Gyroporus cyanescens	[edible]
Hydnum repandum	[edible]
Hygrophorus hypothejus	[edible]
Hypholoma capnoides	[edible]
Hypholoma epixanthum	[edible]
Kuehneromyces mutabilis	[edible]
Laccaria laccata	[edible]
Lactarius acris	[edible]
Lactarius controversus	[edible]
Lactarius deliciosus	[edible]
Lactarius glyciosmus	[edible]
Lactarius helvus	[edible]
Lactarius insulsus	[edible]
Lactarius lignyotus	[edible]
Lactarius necator	[edible]
Lactarius pallidus	[edible]
Lactarius piperatus	[edible]
Lactarius porninsis	[edible]
Lactarius quietus	[edible]
Lactarius repraesentaneus	[edible]
Lactarius resimus	[edible]
Lactarius rufus	[edible]
Lactarius sanguifluus	[edible]
Lactarius scrobiculatus	[edible]
Lactarius semisanguifluus	[edible]
Lactarius subdulcis	[edible]
Lactarius torminosus	[edible]
Lactarius vellereus	[edible]
Lactarius vietus	[edible]
Lactarius violascens	[edible]
Lactarius volemus	[edible]
Lactarius zonarius	[edible]
Langermannia gigantea	[edible]
Lepiota lilacea	[edible]
Lepista irina	[edible]
Lepista nuda	[edible]
Leucopaxillus giganteus	[edible]
Lycoperdon perlatum	[edible]
Lyophyllum decastes	[edible]
Macrolepiota excoriata	[edible]
Macrolepiota procera	[edible]
Marasmius alliaceus	[edible]
Marasmius oreades	[edible]

Marasmius prasiosmus	[edible]
Marasmius scorodonius	[edible]
Morchella esculenta	[edible]
Paxillus atrotomentosus	[edible]
Paxillus involutus	[edible]
Pholiota squarrosa	[edible]
Pleurotus ostreatus	[edible]
Pluteus cervinus	[edible]
Porphyrellus pseudoscaber	[edible]
Ramaria mairei	[edible]
Rozites caperatus	[edible]
Russula adusta	[edible]
Russula aeruginea	[edible]
Russula alutacea	[edible]
Russula atropurpurea	[edible]
Russula aurata	[edible]
Russula badia	[edible]
Russula brunneoviolacea	[edible]
Russula caerulea	[edible]
Russula claroflava	[edible]
Russula cyanoxantha	[edible]
Russula decolorans	[edible]
Russula delica	[edible]
Russula emetica	[edible]
Russula farinipes	[edible]
Russula fellea	[edible]
Russula firmula	[edible]
Russula foetens	[edible]
Russula heterophylla	[edible]
Russula integra	[edible]
Russula lepida	[edible]
Russula maculata	[edible]
Russula melliolens	[edible]
Russula mustelina	[edible]
Russula nigricans	[edible]
Russula ochroleuca	[edible]
Russula paludosa	[edible]
Russula pectinata	[edible]
Russula rosea	[edible]
Russula sanguinea	[edible]
Russula sardonia	[edible]
Russula vesca	[edible]
Russula virescens	[edible]
Russula xerampelina	[edible]
Sarcodon imbricatus	[edible]
Scleroderma aurantiacum	[edible]
Scutiger ovinus	[edible]
Sparassis crispa	[edible]
Strobilomyces floccopus	[edible]
Suillus bovinus	[edible]
Suillus cavipes	[edible]
Suillus granulatus	[edible]
Tricholoma flavovirens	[edible]
Tricholoma imbricatum	[edible]
Tricholoma populinum	[edible]
Tricholoma portentosum	[edible]
Tricholoma robustum	[edible]
Tricholoma saponaceum	[edible]
Tricholoma terreum	[edible]
Tricholomopsis rutilans	[edible]
Tuber aestivum	[edible]
Tylopilus felleus	[edible]
Volvariella bombycina	[edible]
Xerocomus badius	[edible]
Xerocomus chrysenteron	[edible]
Xerocomus parasiticus	[edible]

URUGUAY
Deschamps, 2002

Gymnopilus spectabilis	food
Lactarius deliciosus	food
Laetiporus sulphureus	food
Rhizopogon luteolus	food
Rhizopogon roseolus	food
Suillus granulatus	food
Tricholoma sulphureus	food

UNITED STATES OF AMERICA
1. Birks, 1991; **2.** Lincoff and Mitchel, 1977; **3.** Singer, 1953; **4.** www.mykoweb.com

Agaricus arvensis	edible (4)
Agaricus augustus	edible (4)
Agaricus benesii	edible (4)
Agaricus bernardii	edible (4)
Agaricus bisporus	edible (4)
Agaricus bitorquis	edible (4)
Agaricus campestris	edible (4)
Agaricus cupreobrunneus	edible (4)
Agaricus fuscofibrillosus	edible (4)
Agaricus fuscovelatus	edible (4)
Agaricus lilaceps	edible (4)
Agaricus pattersonae	edible (4)
Agaricus perobscurus	edible (4)
Agaricus silvicola	edible (4)
Agaricus subrutilescens	edible (4)
Aleuria aurantia	edible (4)
Amanita calyptrata	edible (4)
Amanita constricta	edible (4)
Amanita pachycolea	edible (4)
Amanita vaginata	edible (4)
Amanita velosa	edible (4)
Armillaria mellea	edible (4)
Armillaria ponderosa	edible (4)
Battarrea phalloides	medicinal (1)
Boletus aereus	edible (4)
Boletus appendiculatus	edible (4)
Boletus edulis	edible (4)
Boletus truncatus	edible (4)
Boletus zelleri	edible (4)
Bovista pila	medicinal (1)
Bovista plumbea	medicinal (1)
Calvatia craniiformis	medicinal (1)
Calvatia cyathiformis	medicinal (1)
Calvatia utriformis	medicinal (1)
Camarophyllus pratensis	edible (4)
Cantharellus cibarius	edible (4)
Cantharellus subalbidus	edible (4)
Cantharellus tubiformis	edible (4)
Chroogomphus vinicolor	edible (4)
Clitopilus prunulus	edible (4)

Coprinus comatus	edible (4)
Craterellus cornucopioides	edible (4)
Entoloma bloxamii	edible (4)
Entoloma madidum	edible (4)
Flammulina velutipes	edible (4)
Floccularia albolanaripes	edible (4)
Geastrum	medicinal (1)
Gomphus clavatus	edible (4)
Helvella lacunosa	edible (4)
Hericium abietis	edible (4)
Hericium erinaceus	edible (4)
Hericium ramosum	edible (4)
Hydnum repandum	edible (4)
Hydnum umbilicatum	edible (4)
Hypsizygus tessulatus	food (3)
Laccaria amethysteo-occidentalis	edible (4)
Lactarius deliciosus	edible (4)
Lactarius rubidus	edible (4)
Lactarius rubrilacteus	edible (4)
Laetiporus sulphureus	edible (4)
Leccinum manzanitae	edible (4)
Leccinum scabrum	edible (4)
Lepista nuda	edible (4)
Leucoagaricus leucothites	edible (4)
Lycoperdon perlatum	edible (4); medicinal (1)
Lycoperdon pyriforme	medicinal (1)
Macrolepiota rhacodes	edible (4)
Marasmius oreades	edible (4)
Morchella deliciosa	edible (4)
Morganella subincarnata	medicinal (1)
Pleurotus ostreatus	edible (4)
Pluteus cervinus	edible (4)
Sarcodon imbricatus	edible (4)
Sparassis crispa	edible (4)
Suillus brevipes	edible (4)
Suillus pungens	edible (4)
Suillus tomentosus	edible (4)
Tricholoma flavovirens	edible (4)
Tricholoma magnivelare	edible (4)
T. pessundatum var. *populinum*	edible (2)
Tulostoma brumale	medicinal (1)
Volvariella speciosa	edible (4)
Xerocomus chrysenteron	edible (4)

Viet Nam
Burkhill, 1935

Amanitina manginiana	food

Yugoslavia (now Serbia and Montenegro)
1. Richardson, 1988; 2. Zaklina, 1998

Boletus	food (2)
Cantharellus cibarius	food (2)
Craterellus cornucopioides	food (2)
Evernia prunastri	other – perfume (1)

Zambia
1. Pegler and Piearce, 1980; 2. Piearce, 1981; 3. Rammeloo and Walleyn, 1993; 4. Walleyn and Rammeloo, 1994

Afroboletus costatisporus	edible (2)
Amanita flammeola	food (1)
Amanita zambiana	food (1)
Cantharellus cibarius	food (1)
Cantharellus densifolius	food (1)
Cantharellus longisporus	food (1)
Cantharellus miniatescens	food (1)
Cantharellus pseudocibarius	food (1)
Lactarius gymnocarpus	food (1)
Lactarius kabansus	food (1)
Lactarius piperatus	food (1)
Lentinus cladopus	edible (3)
Macrolepiota procera	food (1)
Polyporus moluccensis	edible (4)
Schizophyllum commune	food (1)
Suillus granulatus	edible (2)
Termitomyces clypeatus	food (1)
Termitomyces eurhizus	food (1)
Termitomyces medius	food (1)
Termitomyces microcarpus	food (1)
Termitomyces schimperi	food (1)
Termitomyces titanicus	food (1)
Vanderbylia ungulata	medicinal (4)

Zimbabwe
Boa *et al.*, 2000

Amanita aurea	food
Amanita loosii	food
Amanita zambiana	food
Cantharellus cibarius	food
Cantharellus congolensis	food
Cantharellus miniatescens	food
Cantharellus symoensii	food
Lactarius kabansus	food
Lycoperdon	food
Russula cellulata	food
Termitomyces clypeatus	food
Termitomyces schimperi	food

ANNEX 3

A global list of wild fungi used as food, said to be edible or with medicinal properties

These records are taken from more than 140 sources, including papers, books, websites and other contacts. Full details are held in a database established by the author. The species names are as they appear in the original publication with the exception of obvious spelling mistakes or where the preferred name has changed (Table 5). For mode of nutrition (saprobic, mycorrhizal etc.) see Chang and Mao (1995); Wang, Buchanan and Hall (2002) lists edible fungi that are mycorrhizal. The mycological literature does not always make it clear whether an "edible" fungus is eaten. There must be a clear report to warrant the description of "food" under the column labelled "use". More species are listed at www.wildusefulfungi.org.

(**m**) medicinal properties

Binomial	Use
Afroboletus costatispora	edible
Afroboletus luteolus	food
Agaricus abruptibulbus	edible
Agaricus amboensis	edible
Agaricus arvensis	food (m)
Agaricus augustus	food
Agaricus benesii	edible
Agaricus bernardii	edible
Agaricus bingensis	edible
Agaricus bisporus	food (m)
Agaricus bisporus var. *albidus*	edible
Agaricus bisporus var. *bisporus*	edible
Agaricus bitorquis	food
Agaricus blazei	edible (m)
Agaricus campestris	food (m)
Agaricus comtulus	food
Agaricus croceolutescens	edible
Agaricus cupreobrunneus	edible
Agaricus endoxanthus	edible
Agaricus erythrotrichus	edible
Agaricus essettei	food
Agaricus fuscofibrillosus	food
Agaricus fuscovelatus	edible
Agaricus gennadii	edible
Agaricus goossensiae	edible
Agaricus impudicus	food
Agaricus lilaceps	edible
Agaricus macrosporus	edible
Agaricus micromegethus	edible
Agaricus nivescens	edible
Agaricus pattersonae	edible
Agaricus perobscurus	edible
Agaricus placomyces	edible
Agaricus rodmani	edible
Agaricus rubellus	edible
Agaricus silvaticus	food
Agaricus silvicola	food
Agaricus squamuliferus var. *caroli*	food
Agaricus subedulis	edible
Agaricus subperonatus	food
Agaricus subrufescens	food
Agaricus subrutilescens	food
Agrocybe aegerita	food
Agrocybe broadwayi	food
Agrocybe cylindracea	edible
Agrocybe farinacea	edible
Agrocybe paludosa	edible
Agrocybe parasitica	edible
Agrocybe pediades	edible
Agrocybe salicacicola	edible
Agrocybe vervacti	edible
Albatrellus confluens	edible (m)
Albatrellus ovinus	food
Aleuria aurantia	edible
Amanita alliodora	medicinal
Amanita aurea	food
Amanita bingensis	edible
Amanita caesarea	food
Amanita caesarea f. sp. americana	food
Amanita caesareoides	edible
Amanita calopus	edible
Amanita calyptrata	edible
Amanita calyptratoides	edible
Amanita calyptroderma	food
Amanita ceciliae	food
Amanita chepangiana	edible

Binomial	Use
Amanita constricta	edible
Amanita crocea	food
Amanita flammeola	food
Amanita flavoconia	food
Amanita flavorubescens	edible
Amanita fulva	food
Amanita gemmata	edible
Amanita goosensiae	edible
Amanita hemibapha	food
Amanita hovae	edible
Amanita inaurata	food
Amanita loosii	food
Amanita muscaria	medicinal
Amanita pachycolea	edible
Amanita perphaea	food
Amanita rhodophylla	edible
Amanita robusta	edible
Amanita rubescens	food
Amanita tanzanica	edible
Amanita tuza	food
Amanita umbonata	food
Amanita vaginata	food
Amanita velosa	edible
Amanita virgineoides	edible
Amanita zambiana	food
Amanitina manginiana	food
Amanitopsis pudica	edible
Amauroderma niger	medicinal
Amauroderma rude	medicinal
Arachnion album	food
Armillaria distans	edible
Armillaria luteovirens	food
Armillaria mellea	food (m)
Armillaria ostoyae	food
Armillaria ponderosa	edible
Armillaria tabescens	food
Aspropaxillus lepistoides	edible
Astraeus hygrometricus	edible (m)
Aureobasidium pullulans var. *pullulans*	medicinal
Auricularia auricula-judae	food (m)
Auricularia delicata	food
Auricularia fuscosuccinea	edible
Auricularia mesenterica	edible (m)
Auricularia polytricha	edible
Auricularia tenuis	edible
Bankera fuligineoalba	edible
Battarea phalloides	medicinal
Battarrea stevenii	medicinal
Bjerkandera fumosa	medicinal
Boletellus ananas	food
Boletellus betula	food
Boletellus emodensis	edible
Boletellus russellii	food
Boletinus asiaticus	edible
Boletinus lakei	edible
Boletinus pinetorum	edible
Boletopis leucomelaena	edible
Boletus aereus	edible
Boletus aestivalis	food
Boletus appendiculatus	edible
Boletus atkinsonii	edible
Boletus barrowsii	edible
Boletus bicoloroides	food
Boletus calopus	edible
Boletus citrifragrans	edible
Boletus edulis	food (m)
Boletus emodensis	edible
Boletus erythropus	food
Boletus erythropus var. *novoguineensis*	edible
Boletus felleus	edible
Boletus frostii	food
Boletus griseus	edible
Boletus loyo	food
Boletus luridiformis	edible
Boletus luridus	edible
Boletus michoacanus	food
Boletus nigroviolaceus	edible
Boletus pinicola	food
Boletus pinophilus	food
Boletus regius	edible
Boletus reticulatus	food
Boletus separans	edible
Boletus speciosus	edible
Boletus truncatus	edible
Boletus variipes	food
Boletus violaceofuscus	edible
Boletus vitellinus	edible
Boletus zelleri	edible
Bondarzewia berkeleyii	edible
Bondarzewia montana	edible
Bovista apedicellata	medicinal
Bovista gigantea	edible
Bovista pila	medicinal
Bovista plumbea	edible (m)
Bovista plumbea var. ovalispora	food
Bovista pusilla	medicinal
Bovistella sinensis	medicinal
Buchwaldoboletus spectabilis	edible
Calocera cornea	edible
Calocera viscosa	edible
Calocybe gambosa	food (m)
Calocybe indica	edible
Calocybe leucocephala	edible
Calvatia bovista	medicinal
Calvatia caelata	edible (m)
Calvatia craniiformis	medicinal
Calvatia cyathiformis	food
Calvatia excipuliformis	edible (m)
Calvatia lilacina	edible
Calvatia purpurea	edible
Calvatia utriformis	edible (m)
Camarophyllus niveus	edible
Camarophyllus pratensis	edible
Camarophyllus subpratensis	edible
Camarophyllus virgineus	edible
Cantharellula umbonata	edible

Binomial	Use
Cantharellus cibarius	food (m)
Cantharellus cibarius var. defibulatus	edible
Cantharellus cibarius var. latifolius	edible
Cantharellus cinereus	edible
Cantharellus cinnabarinus	edible
Cantharellus congolensis	food
Cantharellus cyanescens	edible
Cantharellus cyanoxanthus	edible
Cantharellus densifolius	food
Cantharellus eucalyptorum	food
Cantharellus floccosus	edible
Cantharellus floridulus	food
Cantharellus formosus	edible
Cantharellus ignicolor	food
Cantharellus incarnatus	edible
Cantharellus infundibuliformis	edible
Cantharellus isabellinus	edible
Cantharellus longisporus	food
Cantharellus luteocomus	edible
Cantharellus luteopunctatus	edible
Cantharellus miniatescens	food
Cantharellus minor	food
Cantharellus odoratus	food
Cantharellus platyphyllus	edible
Cantharellus pseudocibarius	food
Cantharellus pseudofriesii	edible
Cantharellus ruber	edible
Cantharellus rufopunctatus	edible
Cantharellus rufopunctatus var. ochraceus	edible
Cantharellus splendens	edible
Cantharellus subalbidus	edible
Cantharellus subcibarius	edible
Cantharellus symoensii	food
Cantharellus tenuis	edible
Cantharellus tubiformis	food
Catathelasma imperiale	edible
Catathelasma ventricosum	food
Cerrena unicolor	medicinal
Cetraria islandica	medicinal
Chalciporus piperatus	edible
Chlorophyllum madagacariense	edible
Chlorophyllum molybdites	edible
Choiromyces aboriginum	food
Choiromyces meandriformis	edible
Chroogomphus jamaicensis	food
Chroogomphus rutilus	food
Chroogomphus vinicolor	food
Cladina stellaris	food
Cladonia spp.	medicinal
Clavaria albiramea	edible
Clavaria aurea	edible
Clavaria coralloides	food
Clavaria purpurea	edible
Clavaria vermicularis	food
Clavariadelphus cokeri	food
Clavariadelphus pistillaris	food
Clavariadelphus sachalinensis	edible
Clavariadelphus truncatus	food

Binomial	Use
Clavariadelphus unicolor	food
Claviceps purpurea	medicinal
Clavicorona pyxidata	food
Clavulina amethystina	edible
Clavulina cinerea	food
Clavulina cristata	food
Clavulina rugosa	edible
Clavulinopsis fusiformis	edible
Clavulinopsis helvola	edible
Clavulinopsis miyabeana	edible
Climacocystis borealis	edible
Clitocybe clavipes	food
Clitocybe geotropa	edible
Clitocybe gibba	food
Clitocybe hypocalamus	food
Clitocybe infundibuliformis	edible
Clitocybe nebularis	food
Clitocybe odora	edible
Clitocybe squamulosa	edible
Clitocybe suaveolens	food
Clitopilus abortivus	medicinal
Clitopilus prunulus	food
Collybia acervata	edible
Collybia anombe	edible
Collybia attenuata	edible
Collybia aurea	edible
Collybia butyracea	food
Collybia confluens	food
Collybia contorta	edible
Collybia distorta	edible
Collybia dryophila	edible
Collybia familia	edible
Collybia oronga	edible
Collybia piperata	edible
Collybia platyphylla	edible
Collybia polyphylla	edible
Collybia pseudocalopus	food
Collybia radicata	edible
Collybia subpruinosa	food
Collybia tamatavae	edible
Coltricia cinnamomea	medicinal
Cookeina sulcipes	edible
Cookeina tricholoma	edible
Coprinus acuminatus	edible
Coprinus africanus	food
Coprinus atramentarius	edible (m)
Coprinus castaneus	edible
Coprinus cinereus	edible
Coprinus comatus	edible
Coprinus disseminatus	edible
Coprinus micaceus	food
Coprinus sterquilinus	edible
Corditubera bovonei	edible
Cordyceps militaris	medicinal
Cordyceps ophioglossoides	medicinal
Cordyceps sinensis	edible (m)
Coriolus consors	medicinal
Cortinarius alboviolaceus	edible
Cortinarius armeniacus	edible

Binomial	Use
Cortinarius armillatus	edible
Cortinarius claricolor var. turmalis	edible
Cortinarius collinitus	edible
Cortinarius elatior	edible
Cortinarius glaucopus	food
Cortinarius largus	edible
Cortinarius orichalceus	edible
Cortinarius praestans	food
Cortinarius prasinus	edible
Cortinarius pseudosalor	edible
Cortinarius purpurascens	edible
Cortinarius rufo-olivaceus	food
Cortinarius tenuipes	edible
Cortinarius variecolor	edible
Cotylidia aurantiaca	edible
Craterellus aureus	edible
Craterellus cornucopioides	food (m)
C. cornucopioides var. cornucopioides	edible
C. cornucopioides var. parvisporus	edible
Craterellus fallax	food
Crepidotus applanatus	edible
Crepidotus mollis	edible
Cronartium conigenum	edible
Cryptoderma citrinum	medicinal
Cryptoporus volvatus	medicinal
Cyathus limbatus	medicinal
Cyathus stercoreus	medicinal
Cymatoderma dendriticum	edible
C. elegans subsp. infundibuliforme	edible
Cystoderma amianthinum	edible
Cystoderma terreii	edible
Cyttaria darwinii	food
Cyttaria espinosae	food
Cyttaria gunnii	food
Cyttaria hariotii	food
Cyttaria hookeri	edible
Dacrymyces palmatus	edible
Dacryopinax spathularia	edible
Daedaleopsis confragosa var. tricolor	medicinal
Daldinia concentrica	medicinal
Dictyophora echinovolvata	edible
Dictyophora indusiata f. lutea	edible
Elaphomyces granulatus	medicinal
Endophyllus yunnanensis	edible
Engleromyces goetzii	medicinal
Enteridium lycoperdon	edible
Entoloma abortivum	food
Entoloma aprilis	edible
Entoloma argyropus	edible
Entoloma bloxami	edible
Entoloma clypeatum	food
Entoloma crassipes	edible
Entoloma madidum	edible
Entoloma microcarpum	edible
Evernia mesomorpha	medicinal
Favolus alveolarus	edible
Favolus brasiliensis	food
Favolus brunneolus	food
Favolus striatulus	food
Favolus tesselatus	food
Fibroporia vaillantii	medicinal
Fistulina hepatica	food (m)
Flammulina velutipes	food (m)
Floccularia albolanaripes	edible
Fomes fomentarius	medicinal
Fomes melanoporus	medicinal
Fomitopsis pinicola	medicinal
Fomitopsis ulmaria	medicinal
Fuligo septica	edible
Galiella javanica	medicinal
Ganoderma applanatum	medicinal
Ganoderma capense	medicinal
Ganoderma curtisii	medicinal
Ganoderma lobatum	medicinal
Ganoderma lucidum	edible (m)
Ganoderma sinense	medicinal
Ganoderma tenue	medicinal
Ganoderma tropicum	medicinal
Ganoderma tsugae	edible (m)
Gastrodia elata	edible
Gautieria mexicana	edible
Geastrum fimbriatum	edible
Geastrum hygrometricum	medicinal
Geastrum saccatum	medicinal
Geastrum triplex	food (m)
Geopora sp.	edible
Gloeoporus conchoides	food
Gloeostereum incarnatum	edible
Gomphidius glutinosus	edible
Gomphidius maculatus	edible
Gomphidius purpurascens	edible
Gomphus clavatus	food
Gomphus floccosus	food
Gomphus kauffmanii	food
Goossensia cibarioides	edible
Grifola frondosa	edible (m)
Grifola gargal	food
Gymnopilus earlei	food
Gymnopilus hispidellus	food
Gyrodon intermedius	food
Gyrodon lividus	edible
Gyrodon merulioides	edible
Gyromitra ambigua	edible
Gyromitra antartica	edible
Gyromitra esculenta	edible
Gyromitra infula	food
Gyromitra ussuriensis	edible
Gyroporus castaneus	edible
Hebeloma fastibile	food
Hebeloma mesophaeum	food
Helvella acetabulum	food
Helvella crispa	food
Helvella elastica	food
Helvella infula	food
Helvella lacunosa	food
Hericium abietis	food

Binomial	Use
Hericium caput-ursi	edible
Hericium clathroides	edible
Hericium coralloides	edible
Hericium erinaceus	food (m)
Hericium flagellum	food
Hericium laciniatum	edible
Hericium ramosum	edible
Heterobasidion annosum	medicinal
Hexagonia apiaria	medicinal
Hirschioporus abietinus	medicinal
Hirschioporus fuscoviolaceus	medicinal
Hohenbuehelia petaloides	edible
Hydnopolyporus fimbriatus	edible
Hydnopolyporus palmatus	food
Hydnotrya tulasnei	edible
Hydnum repandum	food
Hydnum umbilicatum	edible
Hygrocybe cantharellus	edible
Hygrocybe coccinea	edible
Hygrocybe conica	edible
Hygrocybe laeta	edible
Hygrocybe nigrescens	food
Hygrocybe obrussea	edible
Hygrocybe psittacina	edible
Hygrocybe punicea	edible
Hygrocybe unguinosa	edible
Hygrophoropsis aurantiaca	food
Hygrophoropsis mangenotii	edible
Hygrophorus agathosmus	edible
Hygrophorus arbustivus	edible
Hygrophorus camarophyllus	edible
Hygrophorus chrysodon	food
Hygrophorus eburneus	edible
Hygrophorus erubescens	edible
Hygrophorus limacinus	edible
Hygrophorus lucorum	edible
Hygrophorus niveus	food
Hygrophorus olivaceoalbus	edible
Hygrophorus penarius	edible
Hygrophorus pudorinus	edible
Hygrophorus purpurascens	food
Hygrophorus russula	food
Hypholoma sublateritium	food
Hypholoma wambensis	edible
Hypomyces lactifluorum	food
Hypomyces macrosporus	edible
Hypsizygus marmoreus	food
Hypsizygus tessulatus	food
Ileodictyon cibarium	edible
Inocybe sp.	edible
Inonotus hispidus	medicinal
Inonotus obliquus	medicinal
Ischnoderma resinosum	medicinal
Kobayasia nipponica	edible
Kuehneromyces mutabilis	edible
Laccaria amethystea	food
Laccaria amethysteo-occidentalis	edible
Laccaria amethystina	food
Laccaria bicolor	food
Laccaria edulis	edible
Laccaria farinacea	edible
Laccaria laccata	food
Laccaria proxima	food
Laccaria scrobiculatus	edible
Laccocephalum mylittae	edible
Lacrymaria velutina	edible
Lactarius akahatsu	food
Lactarius angustus	edible
Lactarius annulatoangustifolius	food
Lactarius camphoratus	edible
Lactarius carbonicola	edible
Lactarius chrysorrheus	edible
Lactarius congolensis	edible
Lactarius controversus	edible
Lactarius corruguis	food
Lactarius deliciosus	food
Lactarius deterrimus	edible
Lactarius denigricans	food
Lactarius densifolius	food
Lactarius edulis	edible
Lactarius flavidulus	edible
Lactarius gymnocarpoides	food
Lactarius gymnocarpus	food
Lactarius hatsudake	food
Lactarius heimii	food
Lactarius indigo	food
Lactarius insulsus	edible
Lactarius inversus	edible
Lactarius japonicus	edible
Lactarius kabansus	food
Lactarius laevigatus	food
Lactarius laeticolor	edible
Lactarius latifolius	edible
Lactarius luteopus	food
Lactarius medusae	food
Lactarius mitissimus	edible
Lactarius necator	edible
Lactarius pelliculatus	edible
Lactarius pelliculatus f. pallidus	edible
Lactarius phlebophyllus	food
Lactarius piperatus	food
Lactarius princeps	edible
Lactarius pseudovolemus	edible
Lactarius pubescens	edible
Lactarius pyrogalus	edible
Lactarius quietus	edible
Lactarius resimus	edible
Lactarius rubidus	edible
Lactarius rubrilacteus	food
Lactarius rubroviolascens	edible
Lactarius rufus	edible
Lactarius salmonicolor	food
Lactarius sanguifluus	edible
Lactarius scrobiculatus	food
Lactarius sesemotani	edible
Lactarius subdulcis	edible
Lactarius subindigo	food
Lactarius tanzanicus	food

Binomial	Use
Lactarius torminosus	edible
Lactarius trivialis	edible
Lactarius vellereus	edible
Lactarius volemoides	food
Lactarius volemus	food (m)
Lactarius xerampelinus	food
Lactarius yazooensis	food
Lactocollybia aequatorialis	food
Laetiporus sulphureus	food
Lampteromyces japonicus	medicinal
Langermannia gigantea	edible (m)
Lanopila nipponica	edible
Lariciformes officianalis	edible (m)
Lasiosphaera fenzlii	medicinal
Leccinum aurantiacum	food
Leccinum chromapes	edible
Leccinum extremiorientale	edible
Leccinum griseum	food
Leccinum manzanitae	edible
Leccinum oxydabile	edible
Leccinum rugosiceps	edible
Leccinum scabrum	food
Leccinum testaceoscabrum	edible
Leccinum versipelle	edible
Lentinellus cochleatus	edible
Lentinula boryana	food
Lentinula edodes	food (m)
Lentinula lateritia	edible
Lentinus araucariae	edible
Lentinus brunneofloccosus	edible
Lentinus critinus	edible
Lentinus cladopus	edible
Lentinus conchatus	edible
Lentinus crinitus	food
Lentinus glabratus	food
Lentinus javanicus	edible
Lentinus praerigidus	food
Lentinus prolifer	edible
Lentinus sajor-caju	edible
Lentinus squarrulosus	food
Lentinus strigosus	food
Lentinus subnudus	edible
Lentinus tigrinus	edible
Lentinus tuber-regium	food (m)
Lentinus velutinus	food (m)
Lenzites betulina	medicinal
Lenzites elegans	edible
Lepiota aspera	edible
Lepiota clypeolaria	edible
Lepiota discipes	edible
Lepiota grassei	edible
Lepiota henningsii	edible
Lepiota madirokelensis	edible
Lepiota mastoidea	edible
Lepiota ventriosospora	edible
Lepista caespitosa	edible
Lepista caffrorum	edible
Lepista glaucocana	edible
Lepista irina	edible
Lepista luscina	edible
Lepista nuda	food (m)
Lepista personata	food
Lepista sordida	edible
Leucoagaricus bisporus	edible
Leucoagaricus hortensis	food
Leucoagaricus leucothites	food
Leucoagaricus rhodecephalus	edible
Leucocoprinus cheimonoceps	food
Leucocoprinus discoideus	edible
Leucocoprinus gandour	edible
Leucocoprinus imerinensis	edible
Leucocoprinus nanianae	edible
Leucocoprinus tanetensis	edible
Leucocortinarius bulbiger	edible
Leucopaxillus giganteus	edible
Lignosus sacer	medicinal
Limacella glioderma	edible
Limacella illinita	edible
Lobaria pulmonaria	medicinal
Lobaria sp.	food
Lycoperdon asperum	medicinal
Lycoperdon candidum	edible
Lycoperdon endotephrum	edible
Lycoperdon gemmatum	edible
Lycoperdon marginatum	edible
Lycoperdon oblongisporum	edible
Lycoperdon peckii	food
Lycoperdon perlatum	food (m)
Lycoperdon pusilum	edible (m)
Lycoperdon pyriforme	food (m)
Lycoperdon rimulatum	edible
Lycoperdon spadiceum	medicinal
Lycoperdon umbrinum	food
Lycoperdon umbrinum var. *floccosum*	edible
Lyophyllum aggregatum	edible
Lyophyllum connatum	edible
Lyophyllum decastes	food (m)
Lyophyllum fumosum	edible
Lyophyllum ovisporum	food
Lyophyllum shimeji	edible
Lyophyllum sykosporum	edible
Lyophyllum ulmarium	edible
Lysurus mokusin	medicinal
Macrocybe gigantea	edible
Macrocybe lobayensis	food
Macrocybe spectabilis	food
Macrolepiota africana	edible
Macrolepiota dolichaula	edible
Macrolepiota excoriata	food
Macrolepiota excoriata var. *rubescens*	edible
Macrolepiota gracilenta	edible
Macrolepiota gracilenta var. *goossensiae*	edible
Macrolepiota procera	food
Macrolepiota procera var. *vezo*	edible
Macrolepiota prominens	edible
Macrolepiota puellaris	edible
Macrolepiota rhacodes	edible
Macrolepiota zeyheri	edible

Binomial	Use
Macropodia macropus	food
Marasmius albogriseus	edible
Marasmius androsaceus	medicinal
Marasmius arborescens	edible
Marasmius buzungolo	edible
Marasmius caryophylleus	edible
Marasmius crinis-equi	edible
Marasmius grandisetulosus	edible
Marasmius heinemannianus	edible
Marasmius hungo	edible
Marasmius maximus	edible
Marasmius oreades	food
Marasmius personatus	edible
Marasmius piperodora	edible
Marasmius purpureostriatus	edible
Marasmius scorodonius	edible
Melanoleuca alboflavida	edible
Melanoleuca brevipes	edible
Melanoleuca evenosa	edible
Melanoleuca grammopodia	edible
Melanoleuca melaleuca	food
Meripilus giganteus	food
Merulius incarnatus	food
Microporus affinis	edible
Microporus xanthopus	medicinal
Micropsalliota brunneosperma	food
Morchella angusticeps	edible
Morchella conica	food
Morchella conica var. rigida	edible
Morchella costata	edible
Morchella crassipes	food
Morchella deliciosa	edible (m)
Morchella elata	food
Morchella esculenta	food (m)
Morchella esculenta var. rotunda	edible
Morchella esculenta var. umbrina	edible
Morchella esculenta var. vulgaris	edible
Morchella intermedia	edible
Morganella subincarnata	medicinal
Mycena aschi	edible
Mycena bipindiensis	edible
Mycena flavescens	edible
Mycena pura	food
Mycenastrum corium	edible
Mycoleptodonoides aitchisonii	edible
Myriosclerotinia caricis-ampullacea	medicinal
Neoclitocybe bissiseda	food
Neolentinus adhaerens	edible
Neolentinus lepideus	edible
Neolentinus ponderosus	food
Nothopanus hygrophanus	edible
Omphalia lapidescens	medicinal
Onnia tomentosa	medicinal
Ophiglossum engelmannii	medicinal
Ossicaulis lignatilis	edible
Otidea onotica	edible
Oudemansiella brunneomarginata	edible
Oudemansiella canarii	food
Oudemansiella mucida	edible
Oudemansiella venoslamellata	edible
Pachyma hoelen	edible
Paecilomyces sinensis	medicinal
Panellus serotinus	edible
Panellus stipticus	medicinal
Panus conchatus	edible
Panus crinitus	edible
Panus flavus	medicinal
Parmelia austrosinensis	food
Parmelia sulcata	medicinal
Paxillus atrotomentosus	edible
Paxillus involutus	edible
Paxina acetabulum	food
Peltigera canina	medicinal
Perenniporia mundula	medicinal
Peziza badia	food
Peziza vesiculosa	edible (m)
Phaeangium lefebvrei	edible
Phaeolepiota aurea	edible
Phaeolus schweinitzii	medicinal
Phaeomarasmius affinis	edible
Phallus fragrans	edible
Phallus impudicus	edible (m)
Phallus indusiatus	medicinal
Phallus tenuis	medicinal
Phellinus rimosus	medicinal
Phellinus baumii	medicinal
Phellinus conchatus	medicinal
Phellinus igniarius	medicinal
Phellinus nigricans	medicinal
Phellorinia inquinans	edible
Phlebopus colossus	food
Phlebopus sudanicus	edible
Pholiota adiposa	edible
Pholiota aurivella	edible
Pholiota austrospumosa	edible
Pholiota bicolor	food
Pholiota edulis	edible
Pholiota highlandensis	edible
Pholiota lenta	food
Pholiota lubrica	edible
Pholiota nameko	edible
Pholiota squarrosa	edible
Phylloporus rhodaxanthus	edible
Picoa carthusiana	edible
Piptoporus betulinus	medicinal
Pisolithus tinctorius	medicinal
Pleurocybella porrigens	edible
Pleurotus abalonus	edible
Pleurotus circinatus	edible
Pleurotus citrinopileatus	edible
Pleurotus concavus	food
Pleurotus cornucopiae	food
Pleurotus cystidiosus	edible
Pleurotus djamor	food
Pleurotus dryinus	food
Pleurotus eryngii	food
Pleurotus eryngii var. ferulae	edible
Pleurotus ferulae	edible

Binomial	Use
Pleurotus flexilis	edible
Pleurotus floridanus	edible
Pleurotus fossulatus	edible
Pleurotus levis	food
Pleurotus nepalensis	edible
Pleurotus ostreatoroseus	edible
Pleurotus ostreatus	food (m)
Pleurotus ostreatus var. magnificus	edible
Pleurotus pulmonarius	edible
Pleurotus rhodophyllus	edible
Pleurotus roseopileatus	edible
Pleurotus salignus	edible
Pleurotus sapidus	edible
Pleurotus smithii	edible
Pleurotus spodoleucus	edible
Pleurotus squarrosulus	food
Plicaria badia	edible
Pluteus aurantiorugosus	food
Pluteus cervinus	food
Pluteus cervinus var. ealaensis	edible
Pluteus coccineus	edible
Pluteus leoninus	edible
Pluteus pellitus	edible
Pluteus subcervinus	edible
Pluteus tricuspidatus	edible
Podabrella microcarpa	edible
Podaxis pistillaris	edible (m)
Podoscypha nitidula	edible
Pogonomyces hydnoides	food
Polyozellus multiplex	edible
Polyporus alveolaris	medicinal
Polyporus aquosus	food
Polyporus arcularius	food
Polyporus badius	edible
Polyporus blanchetianus	edible
Polyporus brasiliensis	edible
Polyporus elegans	medicinal
Polyporus grammocephalus	food
Polyporus indigenus	food
Polyporus moluccensis	edible
Polyporus mylittae	food (m)
Polyporus rugulosus	medicinal
Polyporus sanguineus	edible
Polyporus sapurema	food
Polyporus squamosus	edible
Polyporus stipitarius	food
Polyporus tenuiculus	edible
Polyporus tinosus	medicinal
Polyporus tricholoma	food
Polyporus tubaeformis	medicinal
Polyporus tuberaster	medicinal
Polyporus umbellatus	edible (m)
Polystictus unicolor	medicinal
Porphyrellus atrobrunneus	edible
Porphyrellus pseudoscaber	edible
Psathyrella atroumbonata	food
Psathyrella candolleana	food
Psathyrella coprinoceps	food
Psathryella hymenocephala	food
Psathyrella pululiformis	edible
Psathyrella rugocephla	edible
Psathyrella spadicea	edible
Pseudocraterellus laeticolor	edible
Pseudohydnum gelatinosum	edible
Psiloboletinus lariceti	edible
Psilocybe spp.	medicinal
Psilocybe zapotecorum	edible
Ptychoverpa bohemica	food
Pulveroboletus aberrans	edible
Pycnoporus cinnabarinus	edible (m)
Pycnoporus coccineus	medicinal
Pycnoporus sanguineus	food (m)
Ramalina ecklonii	edible
Ramaria apiculata	edible
Ramaria araiospora	food
Ramaria aurea	food
Ramaria bonii	edible
Ramaria botrytis	food
Ramaria botrytoides	edible
Ramaria cystidiophora	edible
Ramaria fistulosa	edible
Ramaria flava	food
Ramaria flavobrunnescens	food
Ramaria flavobrunnescens var. aurea	food
Ramaria formosa	edible
Ramaria fuscobrunnea	food
Ramaria obtusissima	food
Ramaria ochracea	edible
Ramaria pulcherrima	edible
Ramaria rosella	edible
Ramaria rubiginosa	food
Ramaria rubripermanens	food
Ramaria sanguinea	food
Ramaria stricta	edible
Ramaria subaurantiaca	food
Ramaria subbotrytis	food
Rhizopogon luteolus	edible
Rhizopogon piceus	edible
Rhizopogon roseolus	edible
Rhizopogon rubescens	edible
Rhodophyllus aprilis	edible
Rhodophyllus clypeatus	food
Rhodophyllus crassipes	edible
Rigidoporus sanguinolentus	medicinal
Rigidoporus ulmarius	medicinal
Rozites caperatus	food
Rubinoboletus luteopurpureus	edible
Russula aciculocystis	edible
Russula adusta	edible
Russula aeruginea	food
Russula afronigricans	edible
Russula albonigra	edible
Russula alutacea	food
Russula amaendum	edible
Russula atropurpurea	edible
Russula atrovirens	edible
Russula aurata	edible
Russula brevipes	food

BINOMIAL	USE
Russula cellulata	food
Russula chamaeleontina	edible
Russula chloroides	edible
Russula ciliata	edible
Russula compressa	edible
Russula congoana	edible
Russula consobrina	edible
Russula cyanoxantha	food
Russula cyclosperma	edible
Russula delica	food
Russula densifolia	food
Russula diffusa var. diffusa	edible
Russula eburneoareolata	edible
Russula emetica	edible
Russula erythropus	edible
Russula flava	edible
Russula foetens	food
Russula fraglIls	edible
Russula heimii	edible
Russula heterophylla	food
Russula hiemisilvae	edible
Russula lepida	food
Russula liberiensis	edible
Russula lutea	food
Russula macropoda	edible
Russula madegassensis	edible
Russula mariae	food
Russula mexicana	edible
Russula minutula	edible
Russula nigricans	food
Russula nitida	edible
Russula ochroleuca	edible
Russula olivacea	food
Russula olivascens	edible
Russula ornaticeps	edible
Russula pectinatoides	edible
Russula phaeocephala	edible
Russula pseudoamaendum	edible
Russula pseudostriatoviridis	edible
Russula punctata	edible
Russula queletii	edible
Russula romagnesiana	food
Russula rosacea	edible
Russula rosea	edible
Russula roseoalba	edible
Russula roseostriata	edible
Russula rubra	edible
Russula rubroalba	edible
Russula sanguinea	food
Russula sardonia	edible
Russula schizoderma	edible
Russula sese	edible
Russula sesenagula	edible
Russula striatoviridis	edible
Russula sublaevis	edible
Russula tanzaniae	edible
Russula vesca	edible
Russula violeipes	food
Russula virescens	food (m)
Russula viscida	edible
Russula xerampelina	food
Sarcodon aspratus	food
Sarcodon imbricatus	food
Sarcodon lobatus	edible
Sarcoscypha coccinea	food
Sarcosphaera eximia	food
Schizophyllum brevilamellatum	edible
Schizophyllum commune	food (m)
Schizophyllum fasciatum	edible
Scleroderma bovonei	edible
Scleroderma citrinum	edible
Scleroderma flavidum	medicinal
Scleroderma radicans	edible
Scleroderma texense	edible
Scleroderma verrucosum	edible (m)
Sclerotium glucanicum	medicinal
Scutiger ovinus	edible
Secotium himalaicum	edible
Secotium sp.	medicinal
Shiraia bambusicola	medicinal
Sparassis crispa	food
Sphaerothallia esculenta	food
Sporisorium cruentum	food
Stereopsis hiscens	edible
Stereum hirsutum	medicinal
Stereum membranaceum	medicinal
Strobilomyces confusus	edible
Strobilomyces coturnix	edible
Strobilomyces floccopus	food
Strobilomyces velutipes	edible
Stropharia coronilla	food
Stropharia rugosoannulata	edible
Suillus abietinus	edible
Suillus acidus	edible
Suillus americanus	food
Suillus bovinus	edible
Suillus brevipes	food
Suillus cavipes	food
Suillus granulatus	food
Suillus grevillei	edible (m)
Suillus hirtellus	food
Suillus lactifluus	edible
Suillus luteus	food (m)
Suillus placidus	edible
Suillus plorans	edible
Suillus pseudobrevipes	food
Suillus pungens	edible
Suillus subluteus	edible
Suillus tomentosus	food
Suillus variegatus	edible
Suillus viscidus	edible
Tephrocybe atrata	edible
Terfezia arenaria	edible
Terfezia boudieri	edible
Terfezia claveryi	edible
Terfezia leonis	edible
Terfezia pfeilii	food
Termitomyces albuminosus	food

Binomial	Use
Termitomyces aurantiacus	food
Termitomyces clypeatus	food
Termitomyces cylindricus	edible
Termitomyces entolomoides	edible
Termitomyces eurhizus	food
Termitomyces fuliginosus	edible
Termitomyces globulus	food
Termitomyces heimii	edible
Termitomyces letestui	food
Termitomyces mammiformis	food
Termitomyces medius	food
Termitomyces microcarpus	food (m)
Termitomyces radicatus	edible
Termitomyces robustus	food
Termitomyces schimperi	food
Termitomyces singidensis	food
Termitomyces striatus	edible
Termitomyces striatus var aurantiacus	edible
Termitomyces titanicus	food
Termitomyces umkowaanii	edible
Thelephora ganbajum	food
Thelephora paraguayensis	medicinal
Tirmania africana	edible
Tirmania nivea	edible
Tirmania pinoyi	edible
Trametes albida	medicinal
Trametes cubensis	food
Trametes hirsuta	medicinal
Trametes ochracea	food
Trametes orientalis	medicinal
Trametes pubescens	medicinal
Trametes robiniophila	edible
Trametes sanguinea	medicinal
Trametes suaveolens	medicinal
Trametes versicolor	edible (m)
Tremella aurantia	edible (m)
Tremella concrescens	edible
Tremella foliacea	edible
Tremella fuciformis	edible (m)
Tremella lutescens	edible
Tremella mesenterica	edible (m)
Tremella reticulata	food
Tremellodendron schweinitzii	edible
Tremiscus helvelloides	edible
Trichaptum trichomallum	food
Tricholoma atrosquamosum	edible
Tricholoma bakamatsutake	edible
Tricholoma caligatum	food
Tricholoma equestre	food
Tricholoma flavovirens	food
Tricholoma fulvum	edible
Tricholoma imbricatum	edible
Tricholoma japonicum	edible
Tricholoma magnivelare	food
Tricholoma matsutake	food (m)
Tricholoma mauritianum	edible
Tricholoma mongolicum	edible
Tricholoma muscarium	edible
Tricholoma orirubens	edible

Binomial	Use
Tricholoma pessundatum	edible
Tricholoma pessundatum var. populinum	edible
Tricholoma populinum	food
Tricholoma portentosum	edible
Tricholoma quercicola	edible
Tricholoma saponaceum	edible
Tricholoma scabrum	edible
Tricholoma sejunctum	food
Tricholoma spectabilis	edible
Tricholoma sulphureum	food
Tricholoma terreum	edible
Tricholoma ustaloides	edible
Tricholoma vaccinum	edible
Tricholomopsis decora	edible
Tricholomopsis rutilans	edible
Trogia infundibuliformis	edible
Tuber aestivum	food
Tuber borchii	food
Tuber brumale	edible
Tuber californicum	edible
Tuber gibbosum	edible
Tuber hiemalbum	edible
Tuber indicum	edible
Tuber magnatum	food
Tuber melanosporum	food
Tuber mesentericum	edible
Tuber moschatum	edible
Tuber oligospermum	edible
Tuber rufum	edible
Tuber sinosum	food
Tubosaeta brunneosetosa	edible
Tulostoma brumale	medicinal
Tylopilus ballouii	edible
Tylopilus felleus	food
Tyromyces sulphureus	medicinal
Umbilicaria esculenta	food (m)
Umbilicaria muehlenbergii	food
Usnea hirta	medicinal
Ustilago esculenta	food (m)
Ustilago maydis	food (m)
Vanderbylia ungulata	medicinal
Vascellum curtisii	edible
Vascellum gudenii	edible
Vascellum intermedium	food
Vascellum pratense	edible
Verpa conica	edible
Volvariella bakeri	edible
Volvariella bombycina	edible
Volvariella diplasia	edible
Volvariella esculenta	food
Volvariella parvispora	edible
Volvariella speciosa	edible
Volvariella terastria	edible
Volvariella volvacea	food (m)
Wolfiporia extensa	edible (m)
Wynnella silvicola	edible
Xanthoconium separans	edible
Xerocomus badius	food
Xerocomus chrysenteron	edible

BINOMIAL	USE
Xerocomus pallidosporus	edible
Xerocomus rubellus	edible
Xerocomus soyeri	edible
Xerocomus spadiceus	edible
Xerocomus subtomentosus	food
Xerocomus versicolor	edible
Xeromphalina campanella	edible (m)
Xerula radicata	medicinal
Xylaria papyrigera	medicinal
Xylaria polymorpha	medicinal
Xylosma flexuosum	edible

ANNEX 4

Edible and medicinal fungi that can be cultivated

This list of 92 names has been prepared from Stamets (2000) and Chang and Mao (1995). The = sign denotes the name as original published and which has since been changed. This list contains only saprobic species and excludes ectomycorrhizal species such as truffles (*Tuber* spp.) that are managed in natural habitats.

Binomial
Agaricus arvensis
Agaricus augustus
Agaricus bisporus
Agaricus bitorquis
Agaricus blazei
Agaricus brunnescens
Agaricus campestris
Agaricus subrufescens
Agrocybe aegerita
Agrocybe cylindracea
Agrocybe molesta
Agrocybe praecox
Albatrellus spp.
Armillaria mellea
Auricularia auricula-judae
Auricularia fuscosuccinea
Auricularia polytricha
Calvatia gigantea
Coprinus comatus
Daedalea quercina
Dictyophora duplicata
Flammulina velutipes
Fomes fomentarius
Ganoderma applanatum
Ganoderma curtisii
Ganoderma lucidum
Ganoderma oregonense
Ganoderma sinense
Ganoderma tenus
Ganoderma tsugae
Grifola frondosa
Hericium coralloides
Hericium erinaceum
Hypholoma capnoides
Hypholoma sublateritium
Hypsizygus marmoreus
Hypsizygus tessulatus
Inonotus obliquus
Kuehneromyces mutabilis
Laetiporus sulphureus
Laricifomes officinalis (= Fomitopsis officinalis)
Lentinula edodes
Lentinus strigosus (=Panus rudis)
Lentinus tigrinus
Lentinus tuber-regium
Lepista nuda
Lepista sordida
Lyophyllum fumosum
Lyophyllum ulmarium (=Hypsizygus ulmarium)
Macrocybe gigantea (=Tricholoma giganteum)
Macrolepiota procera
Marasmius oreades
Morchella angusticeps
Morchella esculenta
Neolentinus lepideus (=Lentinus lepidus)
Oligoporus spp.
Oudemansiella radicata
Oxyporus nobilissimus
Panellus serotinus (=Hohenbuehelia serotina)
Paneolus subalteatus
Paneolus tropicalis
Phallus impudicus
Phellinus spp.
Pholiota nameko
Piptoporus betulinus
Piptoporus indigenus
Pleurocybella porrigens
Pleurotus citrinopileatus
Pleurotus cornucopiae
Pleurotus cystidiosus
Pleurotus djamour
Pleurotus eryngii
Pleurotus euosmus
Pleurotus ostreatus
Pleurotus pulmonarius
Pleurotus rhodophyllus
Pluteus cervinus
Polyporus indigenus
Polyporus saporema
Polyporus umbellatus (= Dendropolyporus umbellatus)
Psilocybe cyanescens
Schizophyllum commune
Sparassis crispa
Stropharia rugusoannulata
Trametes cinnabarinum
Trametes versicolor
Tremella fuciformis
Volvariella bombacyina
Volvariella volvacea
V. volvacea var. *gloiocephala*

ANNEX 5

Wild edible fungi sold in local markets

The following examples are mostly from developing countries. It is a small selection of the many species that are sold around the world, particularly for China. Popular species such as *Boletus edulis, Cantharellus cibarius* and *Pleurotus ostreatus* are sold in many countries and are not listed below. Species sold in Malawi or Mozambique markets are available separately (www.malawifungi.org). There are markets for edible fungi in the United Republic of Tanzania (Härkönen, 1995) and Burundi (Buyck, 1994b) but further information is needed on the species sold. Some market reports list only local names.

* indicates species that are also cultivated; it is not always made clear what origin these have in some markets.

ARMENIA Nanaguylan, 2002, personal communication
Agaricus campestris
Agaricus silvaticus
Armillaria mellea
Calocybe gambosa
Cantharellus cibarius
Lactarius deliciosus
Lepista nuda
Lepista personata
Macrolepiota excoriata
Macrolepiota procera
Pleurotus eryngii
Suillus granulatus
Suillus luteus

BOLIVIA Boa, 2001, personal communication
Leucoagaricus hortensis

CHILE Minter, 2002, personal communication
Cyttaria espinosae

CHINA Chamberlain, 1996; Härkönen, 2000; Priest, 2002, personal communication; Winkler, 2002
*Agaricus blazei**
*Auricularia auricula-judae**
Boletus (in the broad sense)
Boletus edulis
*Cordyceps sinensis**
*Dictyophora indusiata**
*Flammulina velutipes**
*Ganoderma lucidum**
*Hericium erinaceus**
Hydnum repandum
Lactarius akahatsu
Lactarius deliciosus
Lactarius hatsudake
Lactarius subindico
Lyophyllum decastes
Pleurotus ostreatus *
Ramaria stricta
Russula spp.
Tricholoma matsutake
Tricholoma quercicola
Umbilicaria esculenta

GUATEMALA Flores, 2002, personal communication
Hypomyces lactifluorum
Ramaria araiospora
Tremella reticulata
Tricholoma flavovirens

INDIA Purkayastha and Chandra, 1985
Coprinus acuminatus
Tricholoma sulphureum

INDONESIA Ducousso, Ba and Thoen, 2002
Scleroderma spp.

KUWAIT
Alsheikh and Trappe, 1983

Tirmania pinoyi

LAO PEOPLE'S DEMOCRATIC REPUBLIC
Hosaka, 2002, personal communication

Amanita hemibapha
Panus rudis
Ramaria sp.
Russula spp.
Schizophyllum commune
Termitomyces sp.

MADAGASCAR
Ducousso, Ba and Thoen, 2002

Cantharellus eucalyptorum

MEXICO
Montoya-Esquivel, 1998; Villarreal and Perez-Moreno, 1989a; www.semarnat.gob.mx

Agaricus campestris
Agaricus silvaticus
Amanita caesarea
Amanita caesarea var. *americana*
Amanita fulva
Amanita rubescens
Amanita tuza
Amanita vaginata
Armillaria mellea
Armillaria ostoyae
Armillaria tabescens
Boletus bicoloroides
Boletus edulis
Boletus frostii
Boletus pinicola
Boletus pinophilus
Boletus reticulatus
Boletus variipes
Calvatia cyathiformis
Cantharellus cibarius
Cantharellus odoratus
Cantharellus tubaeformis
Chroogomphus jamaicensis
Chroogomphus rutilus
Chroogomphus vinicolor
Clavariadelphus truncatus
Clavicorona pyxidata
Clavulina cinerea
Clitocybe clavipes
Clitocybe gibba
Collybia dryophila
Cortinarius glaucopus
Craterellus cornucopioides
Craterellus fallax
Entoloma clypeatum
Gomphus clavatus
Gomphus floccosus
Gomphus kauffmanii
Gyromitra infula
Hebeloma fastibile
Hebeloma mesophaeum
Helvella acetabula
Helvella crispa
Helvella elastica
Helvella infula
Helvella lacunosa
Hygrocybe nigrescens
Hygrophoropsis aurantiaca
Hygrophorus chrysodon
Hygrophorus niveus
Hygrophorus russula
Hypomyces lactifluorum
Laccaria amethystina
Laccaria bicolor
Laccaria laccata
Lactarius deliciosus
Lactarius indigo
Lactarius salmonicolor
Lactarius yazooensis
Laetiporus sulphureus
Leccinum aurantiacum
Lentinula boryana
Lepista nuda
Lycoperdon perlatum
Lycoperdon pyriforme
Lyophyllum decastes
Lyophyllum ovisporum
Marasmius oreades
Morchella conica
Morchella crassipes
Morchella elata
Morchella esculenta
Paxina acetabulum
Pholiota lenta
Pluteus aurantiorugosus
Pluteus cervinus
Ramaria aurea
Ramaria botrytis
Ramaria flavobrunnescens
Ramaria rubiginosa
Ramaria rubripermanens
Rhodophyllus abortivus (Entoloma abortivum?)
Rozites caperatus
Russula alutacea
Russula brevipes
Russula cyanoxantha
Russula delica
Russula mariae
Russula olivacea
Russula romagnesiana
Russula xerampelina
Sarcodon imbricatus
Sarcosphaera eximia
Sparassis crispa
Stropharia coronilla
Suillus americanus
Suillus brevipes
Suillus cavipes
Suillus granulatus

Suillus luteus
Suillus pseudobrevipes
Tricholoma flavovirens
Tricholoma magnivelare
Tylopilus felleus
Ustilago maydis

NEPAL
Adhikari, 1999; Adhikari and Durrieu, 1996

Cantharellus cibarius
Clavulina cinerea
Clavulina cristata
Craterellus cornucopioides
Grifola frondosa
*Hericium erinaceus**
*Hericium flagellum**
Hydnum repandum
Laccaria amethystina
Laccaria laccata
Laetiporus sulphureus
Meripilus giganteus
Pluteus cervinus
Polyporus arcularius
Ramaria aurea
Ramaria botrytis
Ramaria flava
Ramaria fuscobrunnea
Ramaria obtusissima
Termitomyces eurhizus

SENEGAL
Ducousso, Ba and Thoen, 2002

Gyrodon intermedius
Phlebopus sudanicus

TAIWAN PROVINCE OF CHINA
Kawagoe, 1924

Ustilago esculenta

TANZANIA [UNITED REPUBLIC OF]
Härkönen, Saarimäki and Mwasumbi 1994a

Lactarius kabansus
Lactarius phlebophyllus
Russula cellulata
Termitomyces letestui
Termitomyces singidensis

THAILAND
Jones, Whalley and Hywel-Jones, 1994

Auricularia sp.
Cantharellus minor
*Lentinula edodes**
Lentinus praerigidus
Russula aeruginea
Russula lepida
Russula sanguinea
Russula violeipes
*Volvariella volvacea**

TURKEY
Sabra and Walter, 2001

Boletus edulis
Cantharellus cibarius
Rhizopogon sp.
Terfezia boudieri

ZAMBIA
Pegler and Piearce, 1980

Amanita zambiana
Cantharellus cibarius
Cantharellus densifolius
Cantharellus longisporus
Cantharellus miniatescens
Cantharellus pseudocibarius
Lactarius kabansus
Schizophyllum commune
Termitomyces clypeatus
Termitomyces microcarpus
Termitomyces titanicus